To David Riddle, Dan Thornbury,
and Craig O'Connell

Digital Music Wars

Critical Media Studies
Institutions, Politics, and Culture

Series Editor
Andrew Calabrese, University of Colorado

Recent Titles in the Series

Changing Concepts of Time
Harold A. Innis
Mass Communication and American Social Thought: Key Texts, 1919–1968
Edited by John Durham Peters and Peter Simonson
Entertaining the Citizen: When Politics and Popular Culture Converge
Liesbet van Zoonen
A Fatal Attraction: Public Television and Politics in Italy
Cinzia Padovani
The Medium and the Magician: Orson Welles, the Radio Years, 1934–1952
Paul Heyer
Contracting Out Hollywood: Runaway Productions and Foreign Location Shooting
Edited by Greg Elmer and Mike Gasher
Global Electioneering: Campaign Consulting, Communications, and Corporate Financing
Gerald Sussman
Democratizing Global Media: One World, Many Struggles
Edited by Robert A. Hackett and Yuezhi Zhao
The Film Studio: Film Production in the Global Economy
Ben Goldsmith and Tom O'Regan
Raymond Williams
Alan O'Connor
Why TV Is Not Our Fault: Television Programming, Viewers, and Who's Really in Control
Eileen R. Meehan
Media, Terrorism, and Theory: A Reader
Edited by Anandam P. Kavoori and Todd Fraley

Digital Music Wars

Ownership and Control of the Celestial Jukebox

Patrick Burkart and Tom McCourt

ROWMAN & LITTLEFIELD PUBLISHERS, INC.
Lanham • Boulder • New York • Toronto • Oxford

ROWMAN & LITTLEFIELD PUBLISHERS, INC.

Published in the United States of America
by Rowman & Littlefield Publishers, Inc.
A wholly owned subsidiary of The Rowman & Littlefield Publishing Group, Inc.
4501 Forbes Boulevard, Suite 200, Lanham, Maryland 20706
www.rowmanlittlefield.com

P.O. Box 317, Oxford OX2 9RU, UK

British Library Cataloguing in Publication Information Available

Library of Congress Cataloging-in-Publication Data
Burkart, Patrick, 1969–
Digital music wars : ownership and control of the celestial jukebox / Patrick Burkart and Tom McCourt.
p. cm. — (Critical media studies)
Includes bibliographical references and index.
ISBN-13: 978-0-7425-3668-5 (cloth : alk. paper)
ISBN-13: 978-0-7425-3669-2 (pbk. : alk. paper)
ISBN-10: 0-7425-3668-8 (cloth : alk. paper)
ISBN-10: 0-7425-3669-6 (pbk. : alk. paper)
1. Music and the Internet. 2. Sound recording industry. 3. Digital communications. I. McCourt, Tom, 1958– II. Title. III. Series.
ML3790.B845 2006
338.4'778—dc22

2005024966

Printed in the United States of America

∞™ The paper used in this publication meets the minimum requirements of American National Standard for Information Sciences—Permanence of Paper for Printed Library Materials, ANSI/NISO Z39.48-1992.

Contents

Tables and Figures

TABLES

FIGURES

Acknowledgments

Miriam Aune, David Bywaters, and Julie McCourt provided invaluable editing help. We gained much from the insights of Rob Drew, Meg Fisher, Arjun Khanna, Cynthia Meyers, Vinnie Mosco, Eric Rothenbuhler, Willard Uncapher, Brock Craft, and Nabeel Zuberi. Gary Burns and Will Kreth provided assistance and information. Funding assistance was provided by the Melbern G. Glasscock Center for Humanities Research at Texas A&M and the McGannon Center at Fordham University. Thanks to Rita Chang, Kristin Hill, and Jessie Walton for research assistance. Francis E. Dec, Esq., provided important legal counsel. Thanks to Cate McCourt for being such a hell spawn. Finally, our great appreciation goes to Gene and Jo Bernofsky, Steve Bozzone, Darren Campeau, Sandy Carter, Ash Corea, John Downing, Jen Evans, Shuchi Kothari, Becky Lentz, Jane Martin, George and Helen McCourt, Julie McCourt, David Riddle, Michael Shirk, Sharon Strover, Olivier Tchouaffe, and Dan Thornbury for tea and sympathy.

1

The "Celestial Jukebox"

Faith in technology is central to American culture. Technological innovation has routinely inspired gaping wonder and pious awe in our country; in "The Dynamo and the Virgin," Henry Adams described his epiphany at the Great Exposition of 1900:

> To Adams, the dynamo became a symbol of infinity. As he grew accustomed to the great gallery of machines, he began to feel the forty-foot dynamos as a moral force, much as the early Christians felt the Cross. . . . Before the end one began to pray to it; inherited instinct taught the natural expression of man before infinite and silent force.[1]

Similar epiphanies have been deployed to welcome the "Celestial Jukebox," our term for the various systems whereby any text, recording, or audiovisual artifact can be made available instantaneously via wired and wireless broadband channels to Internet appliances or home computers. Futurists regard this as the preeminent model for the future distribution of digital media,[2] which will bring about a kind of technological utopia. Creators of intellectual property will regain control over copyright while liberating themselves from barriers to entry and the interference of distributors. Distributors will gain a huge new revenue stream while eliminating material costs, overhead, and geographic boundaries. Consumer electronics and computer companies will profit from the sale of new recorders, playback systems, and auxiliary devices; technology companies from patents on anticopying software and license fees; and service providers (like telephone and cable companies) from growing demand for lucrative broadband services. And, finally, consumers will be given innumerable choices at low cost as the Internet becomes a "vast intellectual commons" in which "nothing will ever again be out of print or impossible to find; every scrap of human culture transcribed, no matter how obscure or commercially unsuccessful, will be available to all."[3]

From the "electronic brain" public-relations campaigns of the 1940s and 1950s to Bill Gates's "frictionless capitalism" or Alvin Toffler's "Third Wave," promoters of new information technologies have relied on catchphrases exhibiting a facile technological determinism. With their "information superhighway" rhetoric, U.S. President Bill Clinton, Vice President Al Gore, and Federal Communications Commission (FCC) Chairman Reed Hundt linked the model of private travel via the U.S. interstate-highway system—a massive infrastructure project launched by Dwight Eisenhower—to home-computer usage and a rush to "get online." Gore promoted modernization of the telecommunication industries as an inspiring chapter in human progress.

> As we enter this new millennium, we are learning a new language. It will be the lingua franca of the new age. It is made up of ones and zeros and bits and bytes. But as we master it . . . as we bring the digital revolution into our homes and schools . . . we will be able to communicate ideas, and information . . . with an ease never before thought possible. We meet today on common ground, not to predict the future but to make firm the arrangements for its arrival. Let us master and develop this new language together. The future really is in our hands.[4]

The passage of the Telecommunications Act of 1996 in the United States brought on another wave of utopian rhetoric from politicians and policymakers eager to justify media deregulation. FCC commissioner Rachelle B. Chong's statement is typical:

> Today's headline is "America enters a new Information Age." Up to now, we've been regulating in a Star Trek era saddled with Gunsmoke style regulations. . . . This bill tears down outdated barriers to competition, and unleashes communications companies to offer exciting new services to Americans. It allows government to get out of the way. . . . Consumers can look forward to lower prices, more choices and inno-

> vative new services, such as tiny pagers, interactive video, and portable satellite telephones that can be used anywhere, anytime. Americans will have a new option of one stop shopping for all their communications needs.[5]

This faith in technology's goodness, this trust in an inevitable and just march through modernity, reflects the utopianism associated with every major technological innovation in communications since the telegraph. Humanist commentary on the Celestial Jukebox tends to make heavy use of this utopianism: Internet breakthroughs, it claims, will end scarcity and thereby resolve political, economic, and social conflicts. Such commentary, however, mystifies and obscures the technocratic ideology that in fact underlies the infrastructure of the Celestial Jukebox.

In media studies, a hodgepodge of literature coalesced in response to the Napster phenomenon, gleefully describing peer-to-peer (P2P) technology's subversion of existing industrial practices and celebrating the Internet as an untrammeled garden of abundance. This literature resembles in many ways the "Madonna studies" in the early 1990s, which saw Madonna in everything pop cultural, and everything pop cultural in Madonna. Madonna was celebrated as subverting dominant paradigms, as trashing cultural taboos. Her tacit support for the music industry's political economy, however, went unnoticed. The superficial transgressiveness admired in both Madonna and Napster created a glare that obscured the less visible economic processes that made them cultural icons.

Technical explanations of the design and implementation of the Celestial Jukebox are no less unreliable than their humanistic counterparts. The various technologies that comprise it seem transparent and innocent when presented separately in the value-neutral language of computer science and engineering. However, combined and evaluated from a historical perspective that takes account of the music industry's unique structure, these technologies can be seen for what they are: an opaque, oppressive, and socially regressive set of institutions for controlling access to popular culture.

The "Jukebox" as Metaphor

Although the provenance of the term "Celestial Jukebox" is uncertain, it received widespread attention with the 1994 publication of Paul Goldstein's *Copyright's Highway: The Law and Lore of Copyright from Gutenberg to the Celestial Jukebox*. Goldstein describes the Jukebox as

> a technology-packed satellite orbiting thousands of miles above the Earth, awaiting a subscriber's order—like a nickel in the old jukebox, and the punch of a button—to connect him to any number of selections from a vast storehouse via a home or office receiver that combines the power of a television set, radio, CD player, VCR, telephone, fax, and personal computer.[6]

The first half of the term implies heavenly attributes, a gift from God (or, perhaps, the Big Four deities of the recording industry), a reward for hard work and good deeds in life, or a down payment on the afterlife. The second half implies the rhetoric of "e-commerce," employed by corporations and technologists eager to "monetize," or derive revenues from, online access to "content" or cultural products.[7] This vision of a techno-cultural revolution, saturated with audiovisual media and permeated with interactivity, still finds currency in the popular and trade press. The Recording Industry Association of America (RIAA), the powerful music-industry trade group and principal agent in the culture industry's technology wars, used the Celestial Jukebox metaphor in 1989 and 1990 in a rhetorical effort to gain legislative support for digital-performance rights on the Internet.[8] More recently, the online music stores themselves adopted the metaphor; the music service Rhapsody/Listen.com, available through Yahoo!'s portal, still calls itself "THE Celestial Jukebox!" on its home page.

Though we adopt the term here, we do so not to endorse but to critique the host of ideological assumptions it implies. We focus not on the spiritual merits or happy abundance enjoyed by the users of the Celestial Jukebox but on the origins and development of its delivery system as part of the recording industry's value chain for production, packaging, and distribution. After many years of legal and technical development, the basis of the Jukebox has been established by the global cultural industries and their technology partners, and it has been accepted by the major stakeholders, including music fans, artists, labels, and legislators. These culture industries have transformed the Internet from a public space into a private distribution platform for media conglomerates. Though online consumers may have greater access to commercial media content, it has come at a cost to their society. Instead of a gateway into a utopian garden of cultural abundance, the Celestial Jukebox has become a tollbooth into a web of privately owned and operated networks where traffic in intellectual property is carefully monitored and controlled, a walled garden of closed networks with restricted access and tightly circumscribed activities.

Components and Implications of the Jukebox

Although the trappings may be new, the pay-per-use business model is grounded in the history of the jukebox. Edison initially assumed that recordings for his "talking machine" would be of the spoken word, and he advertised applications for stenography and speech therapy,[9] but he and others soon realized that the market for recorded musical performances was quite lucrative. Once the phonograph was used for mechanical musical performances, the recorded popular-music market opened.[10] The jukebox, a fixed device that delivered recordings on demand, was one of the first adaptations of the phonograph and dates to 1889. The earliest mass-produced jukeboxes, patented by Louis Glass,[11] were also known as "nickel-in-the-slot machines." The jukebox was first installed at the Palais Royale Saloon in San Francisco, where music was piped to listeners through tubes linked to the device. A prototype of the Celestial Jukebox, involving network architecture and switched telecommunications links, was operating by 1906. Thaddeus Cahill's Telharmonium, also known as the "electrophone" and the "dynamophone,"[12] delivered live organ performances over telephone lines to homes and offices for a subscription. The Telharmonium, hailed as "the first electronic musical instrument,"[13] weighed two hundred tons and required thirty railroad flatcars for transportation. However, the venture failed due to a lack of venture capital and the discovery that "the machine interfered seriously with local telephone calls."[14]

Unlike the lone entrepreneur behind the Telharmonium, the prime movers behind the Celestial Jukebox are a shape-shifting coalition of corporate interests. While often working at cross-purposes, this coalition has attempted to unify behind a technological "fix" for economically moribund entertainment conglomerates. It has worked to create a "pay-per society"[15] in which each cultural expression or piece of information has a price. Through the Jukebox, the industry hopes to enjoy closer, even "one-to-one" relationships with its audiences, while at the same time denying them important choices that could enhance privacy and competition. This Jukebox may not accept your loose change, but it will take your credit card number, all the while collecting and selling information on your habits, preferences, and identity.

As we see it, two technologies figure prominently in any configuration of the Celestial Jukebox. First, customer-relationship management (CRM) technologies, based on profiles of user attributes and behavior, send personalized content to consumers while assembling merchandising dossiers that can be used in-house to automate marketing campaigns or can be sold to outside interests. Second, digital-rights management (DRM) technologies require that consumers access this content exclusively on the provider's terms, as stipulated in user licensing agreements, terms of service, "click-wrap" agreements, and Federal law. These intermeshing technologies are already combining to extend the existing oligopoly of content distribution and manufacturing power into the digital domain. While the Celestial Jukebox is depicted in the popular press as a perfect, transparent e-commerce marketplace that equally serves producers, distributors, and consumers, it is in fact set to create new boundaries on intellectual property, new restrictions on media access and use, and new threats to individual privacy.

At present, the Celestial Jukebox is being pioneered through online services in which the four major record companies (Universal, Sony BMG, Warner, and EMI, who collectively control 85 percent of global recording sales) provide content and/or hold equity stakes. The Big Four (which were the Big Five before the Sony-BMG merger in 2004) initially resisted digital delivery, adhering to a longstanding pattern in which media industries resist new technologies that threaten to disrupt their organization and practices. The film industry regarded the home-video recorder with considerable suspicion, suing one of its makers in a vain effort to eliminate its use; today, video and DVD releases represent about half of the film industry's revenues. Cable TV once threatened over-the-air television, but many of the would-be competitors eventually merged, and both industries now enjoy substantial profit margins in vertically integrated media enterprises.

Historically, new communication media do not completely and irreparably supplant the formats they follow without adopting some shared or overlapping modes of consumption and distribution.

While digital configurations may allow for new means of distributing text, audio, and visual content, they do not—and, we argue, cannot—fundamentally alter the music industry, which, in the consolidation of record companies in the 1980s and the more recent partial decoupling of major labels from integrated media conglomerates, follows its own imperatives.[16] The Jukebox will not steamroll the music industry or leave record labels defunct once and for all. The music-service providers and clearinghouses that currently make up the Jukebox (described in chapter 4) may very well be transitory and specific to an era of higher broadband Internet availability and music-industry consolidation. However, barring legislation or legal decisions that force some alternative—such as a compulsory licensing scheme—to Jukebox subscriptions or pay-per use, the technological fix for the problem of free music through file sharing will prove irresistible.

CRM technologies will, their promoters claim, enable producers to "know" Internet audiences in much the same way that Nielsen and ratings companies claim to "know" audiences for broadcasting and cable programming. At the same time, DRM technologies will place new and unprecedented limitations on content use and so exert direct control over audiences by restricting the terms and conditions of their access to online content. These intertwined technologies, still intrinsic to the emerging Celestial Jukebox, benefit producers and distributors by reducing their marketing uncertainty, extending their intellectual-property rights, and developing new revenue streams. But they infringe on citizens' rights of privacy, fair use, freedom of expression, and their right to be free from abusive market power.

Legal Underpinnings of the Celestial Jukebox

Coming as it does from the United States, a culture that has been criticized both for worshipping technological idols and for imposing a puritanical cultural homogeneity, the Celestial Jukebox also pushes American-branded intellectual property for a largely American customer base. The vast extension of intellectual-property rights in the United States through the Digital Millennium Copyright Act (DMCA) of 1998 and the Sonny Bono Copyright Term Extension Act of 1998 improved the content industry's competitive advantages in the commercial distribution of digital media. Internationally, the extension of U.S. intellectual-property rights is challenged by, for example, the European tradition of moral rights for artists, which is not recognized in the United States, and European protections for consumer privacy. The benefits afforded to the industry through the DMCA and the Sonny Bono Copyright Term Extension Act do not yet fully extend to copyright holders in the rest of the world. Moreover, there are still many places where a lack of technological infrastructure and disposable income, together with lingering affinities for local cultural practices and artifacts, inhibits the growth of a Celestial Jukebox.

However, intellectual-property protections, and the treatment of purported infringers, are increasingly coordinated internationally. Much of the coordinating effort has been undertaken by the World Intellectual Property Organization (WIPO), a World Trade Organization (WTO) subgroup that describes itself as "an international organization dedicated to promoting the use and protection of works of the human spirit."[17] Multilateral trade agreements in the 1990s, such as those stemming from the WTO and the General Agreement on Tariffs and Trade (GATT), have made the local police forces in member countries into instruments of intellectual-property-rights holders that can pursue criminal sanctions against infringing parties, including consumers, across the globe.

Popular opposition to the extension and deepening of intellectual-property rights in the United States has been relatively weak, though the recording industry's enforcement methods have come under criticism even by mainstream news outlets. Most of the opposition has come after the fact, after intellectual-property laws have diminished fair-use rights and despoiled the Internet's intellectual commons. The RIAA was given the power to craft the language of the DMCA, and it now leads a new intellectual-property enforcement regime. Moreover, in order to make people believe that culture is intellectual property above all else, the RIAA and the Motion Picture Association of America (MPAA) have engaged in a relentless public-relations campaign to convince citizens to respect the principles of the DMCA. In these cynical civics lessons, the RIAA portrays the new rules—though they eliminate key rights protections widely enjoyed before the DMCA—as mere legal expressions of core property values.

The recording industry came late to e-commerce amid a substantial antitrust settlement and a prolonged economic recession. Rather than recognize established Internet culture, with its radical new potential for online distribution of digital media, as an opportunity, the industry attacked it as a threat, constructing, through litigation and lobbying, a legal infrastructure that might enable a generalized technological lockdown of digital "assets" such as music files, digital photographs, e-books, and PDF files. It devoted itself not to creating new models of commerce but to imposing and enforcing new laws to protect itself. In the bread-and-butter customer base to which it might have sought to appeal, it saw not potential consumers but potential criminals.

More importantly, the collective demands of the RIAA and the MPAA for police enforcement of the new digital regime have imposed a surveillance-based governance of the Internet that destroys freedoms for which the Internet is routinely celebrated. We will argue that under the current Jukebox system, the extension of new rights to intellectual-property owners does not preserve or enhance the rights of citizens but in fact interferes significantly with their ability to enjoy the benefits of culture and of a "knowledge-based" economy. It denies rights of free speech and access to information that have been enjoyed for decades in the pre-file-sharing era. Moreover, new intellectual-property laws have severely constricted the ability of the social system to evaluate and correct the problems with the laws themselves.[18] New speech restrictions on researchers and scholars who wish to gather data and publish about DRM, and thereby evaluate the legal infrastructure, take essential tools away from the very people who might provide a corrective to the excesses of the new system.

PLACING DEVELOPMENTS IN PERSPECTIVE

The neoliberal perspective, in which private initiatives are portrayed as the most plausible sources of public good, has dominated legal and legislative discourse regarding the Internet for the last decade. Commercial activity on the Internet was outlawed in the United States until 1991,[19] and the Internet backbone was operated under U.S. government contract until the privatization of the National Science Foundation Network (NSFNet) in 1995.[20] At the same time, a boom in high-tech products and services helped the United States recover from an economic recession. Firms in technopoles like

Boston; New York; Washington, D.C., and Virginia; the North Carolina Research Triangle; Austin; and Silicon Valley boosted computer and information-technology exports, and information-intensive industries contributed increasingly to the U.S. gross national product. Corporations upgraded their computer hardware and software and established presences in cyberspace, and consumers similarly plugged into the Internet en masse. Between 1997 and 2000, the number of users online in the United States rose from approximately 58 million to 120 million.[21]

Neoliberalism and Privatization

The neoliberal aura surrounding the high-tech boom was reflected in the rhetoric of statements by the White House and the FCC, which presented a Celestial Jukebox based on a private, rather than public, initiative as both desirable and inevitable.[22] This rhetoric was intended to rationalize an intense deregulatory push that culminated in the Telecommunications Act of 1996, which removed barriers to mergers and acquisitions in the telecommunications sector and raised barriers to entry. Primary stakeholders, including companies that increasingly consolidated content with distribution, rallied behind Vice President Gore's "National Information Infrastructure Initiative," which promised to unroll a media-saturated carpet across all communities in the country.[23] This initiative collected information from telecommunications, media, and networking companies about the technical and commercial requirements for developing a general-purpose digital-networking platform, and the results were used to promote a deregulated and privatized vision of the Internet to the U.S. Congress.

The privatization of the Internet backbone (a key tenet of proposals for the National Information Infrastructure) increased the Internet's potential for creating markets in media services. Client-server architectures for online transactions proliferated. Portals such as Yahoo!, Excite, Lycos, and America Online (AOL) commercialized the Internet's anarchic public arena through user registrations, search functions, chat communities, and top-down topic-navigation "channels." Portals aggregated traffic and handed online session information off to advertisers like DoubleClick, which profiled Internet users and targeted them with advertisements (in April 2002, DoubleClick settled a privacy lawsuit related to its collection and use of personal user data for $1.8 million). Registration policies allowed the portal networks to profile the identities of users, to merge and enrich these profiles with extra user information, and to establish brand loyalties that would encourage these users to return repeatedly for products and services. As the Internet became increasingly commercialized in the late 1990s, a succession of software companies undertook the technical development of the Celestial Jukebox in hopes of reaping substantial license fees from online merchants, media conglomerates, and other copyright owners. These software firms developed a host of autorecommendation, personalization, and rights-management platforms for online marketing and e-commerce.

Importantly, these technical developments for Internet commercialization occurred in tandem with revisions of intellectual-property law that benefited commercial intellectual-property owners at the expense of society. Although the proponents of the 1996 act claimed that the development of cyberspace would result in increased democratic participation, the public was not invited to express its views to legislators as they created the intellectual-property law that would underpin the Celestial Jukebox.[24] The "pay-per society" expressed itself in the DMCA, which ignored the fair-use precedent established in *Universal v. Sony* and criminalized previously legal computer file sharing by private users, charging for new digital copies instead. Before the Internet was commercialized, peer-to-peer networking was the norm, not the exception, for the flow of online information. However, media conglomerates feared that Napster and peer-to-peer distribution systems would free the Celestial Jukebox from the lockbox they envisioned. The Big Five record companies successfully addressed this threat in court, removing the legal basis for a decentralized alternative to the Celestial Jukebox. Meanwhile, the Big Five—or, now, the Big Four—reinforced their existing market oligopoly through collusive online-distribution arrangements, vertical integration, and mergers and acquisitions.[25]

The neoliberal perspective that laid the groundwork for the Internet's commercialization equated the public interest and competition. By promoting self-regulation among oligopolists, this perspective has had a profound impact on issues of privacy, free speech, fair use, and antitrust. It has enabled the formation of gateway controls on access to online content and licensing. These controls have enabled content providers like the Big Four to search for new marketing venues while neglecting the privacy rights of users, squelching noninfringing uses of copyrighted material, and restricting access by alternative services and independent artists.

Structure of the Book

This book concerns new restrictions—technological, legal, and organizational—on the flow of cultural commodities, imposed by the new political economy of the Celestial Jukebox. These restrictions have angered and confused consumers, for whose protection an incipient social movement is organizing belatedly. As one observer noted,

> We have to remind legislators that intellectual property rights are a socially-conferred privilege rather than an inalienable right, that copying is not always evil (and in some cases is actually socially beneficial) and that there is a huge difference between wholesale "piracy"—the mass-production and sale of illegal copies of protected works—and the file-sharing that most internet users go in for.[26]

Sharing information in collaborative situations produces social benefits in the form of knowledge and culture: creation does not occur in a vacuum. Although we do not entirely exclude reception, our approach is oriented toward production and distribution within the recording industry as representative of global culture industries. In chapter 2, we examine the current structure of the recording industry, a global but U.S.-centric oligopoly of media conglomerates. We identify the possibilities and pitfalls of this conglomerate and oligopolistic structure through a close examination of inter- and intraorganizational conflicts and collaborations over markets, technology, and intellectual property. In chapter 3, we discuss "disruptive" technologies, specifically the MP3 standard for digital files, and the "gift economy"[27] that developed among peers on the Internet and culminated in Napster and its descendents. We examine how the recording industry has attempted to thwart the development of this "gift economy" through legal action while at the same time appropriating elements of it for the Celestial Jukebox. In chapter 4, we examine the rivalry between the "Darknet" and trusted computing, the development of the music-service-provider (MSP) and clearinghouse models for digital distribution, and the components of the Celestial Jukebox: customer-relationship management and digital-rights management. These technologies, when integrated, provide opportunities for surveillance and control of consumer behavior. In chapter 5, we conclude with a stakeholder analysis that considers the outcomes of the establishment of the Celestial Jukebox for consumers, citizens, the entertainment industry, and the state. We also identify the prime movers of the media-reform project in the United States.

In expanding from selling hard goods in stores to selling services via the Jukebox, the recording industry has established a model for control over digital distribution and access. Through technological and legal lockdowns on digital assets in cyberspace, the recording industry and other intellectual-property owners hope to dictate the terms of consumer rights to cultural and information access, even in areas with established consumer rights of fair use and privacy. The social problem we address in this book is the movement from copyright minimalism to copyright maximalism,[28] which deprives consumers of the benefits of knowledge and culture without offering new rights, access, or empowering technologies in return. If "culture" traditionally belongs to a community as a public good, this concept of culture is at risk in the privatized cultural commons of a Celestial Jukebox. Although the Jukebox has been enveloped in an aura of inevitability, we find that technology is a contested area that is defined by social action. The Celestial Jukebox is not inevitable.

NOTES

1. Henry Adams, *The Education of Henry Adams* (1907; repr., New York: Vintage Books, 1990), 353.
2. Paul Goldstein, *Copyright's Highway: The Law and Lore of Copyright from Gutenberg to the Celestial Jukebox* (New York: Hill and Wang, 1994); Charles Mann, "The Heavenly Jukebox," *The Atlantic Monthly*, Sept. 2000; Steve Jones, "Music That Moves: Popular Music, Distribution, and Network Technologies," *Cultural Studies* 16, no. 2 (2002): 213–32.
3. Mann, "Heavenly Jukebox," 41.
4. Al Gore, "Remarks Prepared for Delivery by Vice President Al Gore," Royce Hall, UCLA, Los Angeles, CA, 11 Jan. 1994, http://www.ibiblio.org/icky/speech2.html.
5. Rachelle B. Chong, "FCC Commissioner Chong Hails Passage of New Telecom Bill," 1 Feb. 1996, http://www.fcc.gov/Speeches/Chong/sprbc602.txt.

6. Goldstein, *Copyright's Highway*, 199.

7. Jim Griffin, "Monetizing Anarchy—Can We Pay for Art without Controlling Art?" *Thefeature.com*, 18 Nov. 2002. http://www.thefeature.com/index.jsp?url=view.jsp%3Fpageid=24637.

8. Seth Greenstein, personal correspondence, 30 Nov. 2003.

9. Brian Winston, *Media Technology and Society: A History: From the Telegraph to the Internet* (New York: Routledge, 1988).

10. Reebee Garofalo, "From Music Publishing to MP3: Music and Industry in the Twentieth Century," *American Music* 17, no. 3 (Fall 1999): 323; Winston, *Media Technology*.

11. *Encyclopaedia Britannica Almanac*, 2003, "Encyclopaedia Britannica's Great Inventions," http://corporate.britannica.com/press/inventions.html.

12. Russell Naughton, "Adventures in Cybersound," http://www.acmi.net.au/AIC/TV_TL_COMP_1.html.

13. Richard Zvonar, "Spatial Music," 1999, http://www.zvonar.com/writing/spatial_music/History.html.

14. Anonymous, "Thaddeus Cahill's 'Dynamophone/Telharmonium,'" http://www.obsolete.com/120_years/machines/telharmonium/.

15. Vincent Mosco, *The Pay-Per Society: Computers and Communication in the Information Age* (Mahwah, NJ: Ablex, 1990).

16. Patrick Burkart, "Loose Integration in Markets for Popular Music," *Popular Music & Society* 28, no. 4 (2005): 489–500.

17. World Intellectual Property Organization, "About WIPO," http://www.wipo.int/about-wipo/en/overview.html.

18. Kembrew McLeod, *Freedom of Expression: Overzealous Copyright Bozos and Other Enemies of Creativity* (New York: Doubleday, 2005); Lawrence Lessig, *The Future of Ideas: The Fate of the Commons in a Connected World* (New York: Random House, 2002); Siva Vaidhyanathan, *Copyrights and Copywrongs: The Rise of Intellectual Property and How It Threatens Creativity* (New York: New York University Press, 2001).

19. John Alderman, *Sonic Boom: Napster, MP3, and the New Pioneers of Music* (Cambridge, MA: Perseus Press, 2001).

20. Winston, *Media Technology*.

21. Bob Tedeschi, "E-Commerce," *Wall Street Journal*, 21 May 2001, C9.

22. Reed Hundt, *You Say You Want a Revolution: A Story of Information Age Politics* (New Haven, CT: Yale University Press, 2000).

23. Patricia Aufderheide, *Communications Policy and the Public Interest: The Telecommunications Act of 1996* (New York: Guilford Press, 1999).

24. William Schaumann, "Copyright Infringment and Peer-to-Peer Technology," *William Mitchell Law Review* 28, no. 3 (2002).

25. Tom McCourt and Patrick Burkart, "When Creators, Corporations, and Consumers Collide: Napster and the Development of Online Music Distribution," *Media, Culture and Society* 25, no. 3 (2003).

26. John Naughton, "Intellectual Property Is Theft. Ideas Are for Sharing," *The Observer*, 9 Feb. 2003, http://observer.guardian.co.uk/business/story/0,6903,891687,00.html.

27. Alderman, *Sonic Boom*.

28. Paula Samuelson, "The Copyright Grab," *Wired*, Jan. 1996, http://www.wired.com/wired/archive/4.01/white.paper_pr.html.

2

The Music Industry in Transition

Since audio recordings are part of nearly all audiovisual products, the recording industry is among the most pervasive and fundamental of the entertainment industries.[1] At the same time, it may be the most volatile. Consumers face a deluge of products (814 million recordings were shipped in 2004),[2] and their interests change quickly: musical genres proliferate and evolve in ways that are anything but well defined, especially given the increasing hybridization of music in the global marketplace. Consumers of music soon tire of formulas; taste is not fixed but restless, subjective, and highly unpredictable. The recording industry is fraught with risk. Therefore, it has taken increasingly draconian measures to manage this risk. In the first chapter, we described how the recording industry's model for a Celestial Jukebox may impair the fair-use, privacy, and free-speech rights of consumer citizens and academic researchers. In this chapter, we examine the recording industry itself: the contracts routinely offered to artists, the nature of the music-industry oligopoly, the increasing integration of media markets, and the implications of collusive behavior resulting from regulatory inaction. We argue that the potential for collusion and abuse requires legislation to protect artists from exploitative contracts, and consumers from invasion of privacy and price gouging in the form of rent seeking.

THE STRUCTURE OF THE RECORDING INDUSTRY

While the recording industry has had its booms and busts, its fortunes have improved on the whole throughout its history. Sales grew rapidly throughout the 1920s, crashed in the 1930s, and grew steadily throughout the 1940s.[3] They flattened in the early 1950s but then, driven by the rock and roll market, grew steadily again (except for a few sporadic dips) from 1955 through 1978,[4] after which they flattened until CDs became popular in the early 1980s. They flattened yet again in the mid-1990s, after the novelty of CDs faded and the major record companies had recycled their most popular catalog holdings. Despite an unsuccessful attempt at similar recycling strategies with new (and inferior) technologies such as MiniDiscs and Digital Compact Cassettes, recording sales surged in the late 1990s, reaching a peak of $13 billion annually in the United States.[5] The RIAA blamed sales declines after 2000 on unauthorized file trading; these claims are examined in chapter 3. It is inarguable, however, that the industry is deeply unsettled at present. Since its infrastructure and practices are based on the production, distribution, and promotion of "hard goods" like CDs, the popularity of music on digital downloads is forcing it to either change or face obsolescence. The following section describes the structure of the recording industry, which has in turn shaped the structure of the Celestial Jukebox.

Artists and Record Companies

Production, distribution, and retailing are primary links in the record industry's value chain, while exhibition (performance) and A&R ("artist and repertoire") are supporting functions.[6] Record companies operate like venture capitalists who can also produce, market, and distribute their goods and services. They locate, finance, and develop artists; they oversee production; they market music through promotion, advertising, and airplay; and they distribute recordings through retail and online services.[7] Significantly, most recordings are owned by the companies that finance and market them rather than by the artists whose names appear on them. The artists agree to this arrangement because they stand to benefit from the record company's recording, distribution, and promotional services.

The company compensates the artist with royalty payments (typically 7 to 12 percent for new artists and 15 percent or higher for "superstars") from sales.[8] However, an artist's first payment on a new project usually comes in the form of an "advance" on future royalties, from which the artist pays the costs of recording, promotional videos, and other services. The artist, therefore, will not earn royalties until enough recordings have been sold to recoup these costs and repay the advance. When a recording fails to break even, its "debt" will be applied to any follow-up recordings the artist makes while under contract. Recording contracts usually consist of a series of one-year agreements in which the artist must deliver a minimum number of recordings (traditionally, enough recordings to comprise one album). The record company holds the right to renew these agreements. Thus, an artist's dependency on the label for funding creates a peculiar work relationship: the artist borrows services from the label and repays it with recording sales at terms set by the label. If the label performs its marketing function poorly, the artist may be doubly penalized, first by not collecting royalties and second by suffering the onus of being "dropped" or fired by the label.

Major labels require exclusivity agreements from their most successful artists, which prevent artists from hedging their bets by working with multiple companies. Artists typically sign multiyear contracts in which the record company retains ownership of the master recording. Most of these contracts are negotiated in California or New York; while New York law has no specific limit, under California law, these contracts are limited to seven years.[9] The "seven-year rule" (California Labor Code Section 2855), the result of a legal battle by the actress Olivia De Havilland to free herself from long-term contracts with studios, granted free agency to actors (and other artists and performers, including recording artists) in 1945 by allowing them to negotiate contracts based on their fair market value for terms limited to seven years. However, recording artists occupy a gray area in this regard, since their contracts are contingent on their delivering a specified number of recordings to their record companies, and these companies have increasingly spread release dates apart to shape demand (a strategy that has particularly irked prolific artists such as Prince).

As it gradually widened the gap between releases, in 1985 the recording industry unsuccessfully tried to stretch the length of contracts from seven to ten years. Their efforts met with greater success two years later, when California law was amended to allow record companies to collect damages for undelivered albums, effectively ending the seven-year rule for recording artists.[10] Since record-company contracts typically call for an artist to release seven albums, and since the contracts increasingly limit these releases to one every two years, the "seven-year" contract often stretches to fourteen years, which is also the average lifespan of an artist's recording career. Artists and legislators have likened the result to "indentured servitude."[11] A lawsuit filed by Courtney Love against Vivendi Universal argued that this arrangement unfairly benefits record companies by stretching periods between artists' releases and then holding them liable for recordings they fail to deliver.[12] Love dropped the suit in September 2002 when she renegotiated her contract, but since then, musicians' groups have lobbied for a California bill that would change the labor code to permit recording artists to work more like free agents and avoid restrictive recording contracts.[13]

Artists have often railed against the injustices perpetrated by the recording industry.[14] In 2002, the *New York Times* reported that 99.99 percent of audits revealed that record companies had underpaid their artists.[15] The Dixie Chicks filed suit against Sony in August 2001, seeking to end their contract, claiming that Sony had "cheated them out of more than $4 million by underreporting sales figures and overcharging them for company services" such as producer fees and advances. They settled their suit against Sony in June 2002, after receiving a $20 million advance and a boost in royalty rates from sales to 20 percent.[16] To settle such high-profile disputes, some record companies renegotiated the contracts of disgruntled stars, first offering large cash advances and higher future royalty rates in exchange for additional albums, and then claiming that the renegotiation constituted a new contract with a new seven-year limit.[17]

The obviously imbalanced power relationship between artists and record companies also is apparent in their respective income potentials. Virtually no recording artists see income from their recordings until the record company has recouped its costs. The artists' potential income increases, however, according to their role in the recording's creation. They earn money from sales of recordings only after the record company's advances have been repaid. However, if they also write their songs, they can begin earning money from the outset from "mechanical rights," based on physical manifestations of the song in a CD, LP, cassette, or sheet music, and "performance rights," based on performances of a song on a record and CD, and on both original and recorded performances on radio, television, jukeboxes, in night clubs and concert halls, and in advertisements.

Anyone who uses a song in a commercial context must pay royalties to publishing organizations such as the American Society of Composers, Artists, and Publishers (ASCAP); Broadcast Music Incorporated (BMI); the Society of European Stage Authors and Composers (SESAC); the British Phonogram Performance Limited (PPL); the Canadian Private Copying Collective (CPCC); and the Mexican Sociedad Mexicana de Productores de Fonogramas, Videogramas y Multimedia, S.G.C. (SOMEXFON). The money collected by these publishing organizations is then distributed to songwriters through a formula that factors in audience size, time of day, and length of composition.[18] These "mechanical" and "performance" royalties are a more reliable source of income than recording royalties, on which record companies retain copyright, even if artists have fully repaid their advances. As one observer noted, "Performers rarely see a penny of CD royalties. Unheralded session musicians and orchestra members, who are paid flat fees, often do better in the end."[19]

The Copyright Fix

The lopsided power relationship between artists and the major labels has been justified by one central claim: the recording industry assumes all the financial risk and loses money on most of its products. Only about one recording in ten breaks even,[20] and record companies recoup their investment on only 5 percent of all new artists they sign.[21] In 1999, only eighty-eight recordings (out of nearly three thousand released on the market) accounted for 25 percent of all record sales.[22] Labels put capital at risk in their fixed costs—for licensing operations, recording studios, manufacturing plants, and promotion offices. Labels also expend variable costs for every new release, which climb with increased levels of business activity, including administrative labor, utilities, and other essentials in the production and distribution process. Together, these costs must be spread across all units sold per release, whether one or one million. If the odds of success are so dismal, how, then, do record companies stay in business at all? The percentage of fixed costs for each recording declines, and the profit margin rises, with each sale. When enough copies have been sold to pay the fixed costs, subsequent sales are almost pure profit.

The many recordings that fail to recoup investments are subsidized by the few that become massive sellers, or hits (a "hit" ranks in the top ten on a *Billboard* chart for a week or longer). One hit album such as *Thriller* (with ninety-one weeks on the chart, thirty-seven weeks at number one, and over 26 million copies sold by May 2002) subsidizes a vast number of financial failures. Thus, industry economics require record companies to act as "hit machines," producing numerous iterations of a music formula to score one or two big hits. Furthermore, record-company catalogs can generate money for decades through reissues, compilations, and licensing. The Eagles' *Greatest Hits*, for example, first released on LP in 1976, has been re-released in myriad formats and have now sold over 27 million copies.[23] These secondary music markets smooth the edges of sales trend lines.

The history of copyright in the recording industry began in the late nineteenth century, when New York City's Tin Pan Alley became the site of consolidation for U.S. music publishing houses and "anticipated many of the practices of the music business in latter years," including "song factories," which produced only popular sheet music.[24] Copyright law did not require manufacturers of player-piano rolls and recordings to pay royalties to publishers, since copyright law defined a "fixed" musical form as one that could be accessed without mediating technologies such as player pianos, phonographs, and jukeboxes. Since Tin Pan Alley's stock in trade was sheet music, publishers initially paid relatively little attention to recordings, but in 1909 its corporate leaders, Victor Herbert and John Philip Sousa, lobbied for a revision to copyright laws to benefit their interests.[25] The ensuing legislation, the Copyright Act of 1909, awarded publishers a royalty of two cents for each "mechanical reproduction" (i.e., record or player-piano roll), in addition to performance royalties. According to Marcus, "Congress, fearing the market power of the dominant piano roll company of the day, instituted a compulsory license that allowed manufacturers to use any musical composition without permission, as long as the 2 cent-per-copy royalty was paid to the copyright holder."[26] The 1909 act enabled publishers to gain a stake in the new media and "opened the door for collaborations between publishers and recording companies which had not existed previously."[27]

Revisions of European copyright law in 1928 introduced "moral rights" for authors that the United States did not adopt.[28] In the

United States, ASCAP served as the primary collection agency for performance and mechanical royalties for the major labels, joined by BMI in 1939. Jukeboxes were exempted from paying performance royalties until 1976, when a U.S. Copyright Act revision created a compulsory license in the form of an annual per-jukebox fee. Negotiations between BMI, ASCAP, SESAC, and the Amusement & Music Operators Association yielded a Jukebox License Agreement on rates, which are currently $364 for the first jukebox and $83 for each additional jukebox.[29] Online subscription services are exempted from the Jukebox License Agreement,[30] although the rationale for this exemption is uncertain, since the U.S. Copyright Office's Copyright Arbitration Royalty Panel (CARP) decided in 2002 that webcasters should pay performance royalties.

Webcasting is a relatively recent addition to the Internet, whose sprawling, centerless architecture made it unsuited to broadcast-style communications. Client-server software packages have, however, been developed to overcome the Internet's "limitations" in this respect, so as to permit "streaming" media from media servers. RealAudio, released in 1995, allowed webcasters to license software for encoding and serving streaming media, and users to "tune in" to "radio" streams. Individual users could use SHOUTcast to stream MP3 audio by 1999. The Internet's open architecture also posed technological obstacles for enforcing private-property rights. Webcasters, who were operating royalty-free online jukeboxes until 2002, shortly thereafter joined the ascendant Napster in opening the online taps of free music, much to the horror of the RIAA. After fierce lobbying, CARP determined royalties of .07 cents per commercial performance for radio streamed to the Internet, and .14 cents per performance for Internet streams—another example of a "copyright grab"[31] in cyberspace.

Although the primary assets of the recording industry have been copyrights throughout its history, the relative importance of copyright applications is subject to change. As Frith states,

> In the music industry itself, a song—the basic musical property—represents "a bundle of rights"; income from the song comes from the exploitation of those rights, and what happened in the 1980s was that some of these (the "secondary rights" [i.e., licensing and copyright fees from other users]) became more profitable, others (the "primary rights" [i.e., selling your own records]) less so.[32]

The Internet now evokes an even more radical shift. In chapter 4, we discuss the industry's current strategies for dealing with that shift through the implementation of online-distribution models, through which the Big Four record companies hope to minimize risk by leasing their content to third parties.

The Recording-Industry Oligopoly

According to Garofalo, the history of the music industry has been marked by three phases dominated by three kinds of companies:

> 1. Music publishing houses, which occupied the power center of the industry when sheet music was the primary vehicle for disseminating popular music [from approximately 1886 to 1910];
> 2. Record companies, which ascended to power as recorded music achieved dominance [approximately 1910 to 1980]; and
> 3. Transnational entertainment corporations, which promote music as an ever-expanding series of "revenue streams"—record sales, advertising revenue, movie tie-ins, streaming audio on the Internet—no longer tied to a particular sound carrier [1980 to present].[33]

The recording industry was founded on patents and their litigation, in which diverse patent holders sought to stake their claims on the nascent gramophone industry in the 1890s and early 1900s. When the dust settled, three corporations (Edison, Columbia, and Victor) owned every significant patent regarding records and phonographs.[34] Although the major firms have shifted over time, and the hegemony of the major companies has been challenged at various moments (such as in the mid-1950s, when majors failed to capitalize on the rock-and-roll boom), the recording industry has effectively functioned throughout its history as an oligopoly. A market structure, based on market share of the participants, can be competitive (five or more key players), oligopolistic (four or fewer), duopolistic (two), or monopolistic. Though the record industry includes hundreds of small independent labels, it is so completely dominated by four companies that it is, in effect, an oligopoly.

Mergers and acquisitions among entertainment companies, or conglomerates with media divisions, became widespread in the mid-1980s[35] as the government lifted bans on media cross-ownership. New economies of scope, by which merged operations could make more profits with fewer resources, and new economies of scale, by which larger markets could be served with only marginally increased expenditures, became irresistible. According to a Paine Weber financial analyst, "The good companies must be integrated"—must, that is, combine diverse media and telecommunication activities, including content and delivery.[36]

Ownership concentration in the recording industry increased accordingly.[37] In 1980, six record companies dominated the recording industry: Capitol/EMI, CBS (purchased by Sony in 1988), MCA, Polygram, RCA (purchased from General Electric by Bertelsmann in 1986), and Warner. MCA was forged from a merger of Decca, MCA, and Universal in the early 1960s; the company acquired ABC Records in 1979, Motown in 1986, and Geffen in 1989; and Seagrams purchased 80 percent of the combined operation (rechristened Universal Music Group) in 1995. The Big Five emerged in December 1998 when Seagrams bought Polygram Records from Philips, folding Polygram into its Universal operation. Recent market share among the major labels is indicated in table 2.1; the Big Five became the Big Four in 2004 when BMG merged with Sony Music, concluding a 50 percent loss of ownership diversity in popular-music producers in only eight years.

The recording industry oligopoly depends on vertical integration, in which a company controls its suppliers and distributors.[38] The Big Four own or are affiliated with music publishing firms, A&R,

Table 2.1. Total Market Share for Music Recordings (in percentages)

	Global		*US*	
	1999	*2002*	*1999*	*2002*
Universal	21.8	25.9	28.1	28.9
Sony	19.0	14.1	19.5	15.6
EMI	12.9	12.0	11.4	8.0
Warner	11.9	11.9	12.8	15.9
BMG	11.9	11.1	15.2	15.0
All others	22.5	25.0	13.0	16.0

Sources: Charles Goldsmith, William Boston, and Martin Peers, "EMI, Bertelsmann Unit End Merger Talks," *Wall Street Journal*, 2 May 2001, 25(A); Charles Goldsmith, Matthew Karnitschnig, Martin Peers, and Bruce Orwall, "As Music Sector's Woes Worsen, Sony and BMG Propose a Merger," *Wall Street Journal*, 7 Nov. 2003, 1(A); Jennifer Ordonez, "U.S. Music Sales Stayed Muted in '02," *Wall Street Journal*, 3 Jan. 2003, 8(A).

manufacturing plants, distribution and promotion operations, marketing, record clubs, record-store chains, and digital Internet music sellers. By controlling each step in the link between artists and audiences, vertically integrated companies gain four advantages over their competitors. First, they increase the number of potential revenue sources. A record company that also owns a publishing company can retain the rights to its recordings and transform copyright fees from cost to profit. Second, they can centralize management, accounting, and other administrative functions and coordinate resources across their subsidiaries. In the first years following the turn of the millennium, the music industry eliminated nearly 20 percent of its workforce.[39] Third, integrated companies can establish release and promotion schedules that maximize the sales performance of their overall holdings rather than just individual releases. This is a primary reason why record companies skirt the seven-year rule. Fourth, by controlling virtually the whole production process, vertically integrated companies can introduce barriers to market entry. They can diminish the access of competitors or would-be competitors to marketing and distribution networks, or to the audiences in the media markets themselves. For example, Warner Music benefits not only from the music of its subsidiary labels, but also from any label that contracts with Warner for distribution, such as Music Master, while at the same time denying others access to their distribution and promotion services.

Although recording technology has become increasingly accessible and CD manufacturing costs have dropped to historic lows, the recording industry still presents enormous barriers to entry. The number of songwriters and artists seeking to create hit records is uncountable, but creating a hit record requires access to national and international markets, since overall demand is uncertain and the majority of releases fail to return their investments. The potential to sell millions of recordings involves huge amounts of capital, complex promotion, and vast distribution. Indeed, the majors have traditionally differed from the independents in that they own their distribution systems, while independents must contract with other companies (usually the majors themselves) for distribution. Independent companies must thus share their profits with the distributor and surrender some control of their own products, which limits their autonomy and potential for innovation.

As we will describe, integration comes at a social cost. Although oligopolies tend to be relatively efficient when providing private goods, they tend not to provide the benefits of competition: low price and high quality. An oligopolistic media industry, with a non-competitive marketplace and a business model based on royalty payments, tends to produce goods of low quality and little diversity. Quality and diversity are necessarily related in media industries, where more diverse products can satisfy a broader range of tastes. Research by Peterson and Berger, Rothenbuhler and Dimmick, and Lopes indicates that the diversity of a music catalog (as indicated by record-chart success) suffers during periods when the industry has been highly consolidated.[40] When control of the industry was less concentrated, in the 1960s and 1970s, popular-music charts featured a greater variety of recordings, faster turnover on the charts, and a greater number of new artists than in the 1980s.

However, more recent concentration within the industry has not, at least in the short term, led to the decline in musical diversity that orthodox media-economics theory would have predicted. By maintaining equity in subsidiary labels, the major companies have been able to maintain diverse product lines.[41] The result is a loosely coupled system in which subsidiary labels enjoy relative autonomy in making creative decisions, while their owners maintain centralized financial control. Artists who achieve initial success through independent labels often jump to major labels in order to enjoy greater promotion and distribution. In the 1990s, the majors diversified into larger and larger swaths of genres by "rolling up," or buying, the entire catalogs of other majors or independent labels outright, or by offering distribution in exchange for equity ownership. Many independent artists unwittingly found themselves on major labels in the 1990s.

Still, whether this system continues will depend on the decisions of a very small number of participants, who will sustain it only so long as they stand to profit from it. Small independent companies, which have been crucial to establishing new music genres since the advent of the recording industry, lack the distribution and promotion resources necessary to create hit records even nationally, much less internationally, on their own. Together, they produce only one-quarter of total worldwide music sales, and their survival depends on the sufferance of the oligopoly, which remains firmly in control.

Tight Diversification: The Music Business in the Entertainment Industry

As an increasingly large share of copyrights is held by an ever smaller number of major record companies, these companies are attempting, as parts of major media conglomerates, to extend the scope and scale of their copyrights. In recent decades, music copyrights became increasingly important to entertainment conglomerates. The acquisition of the copyrights, through the acquisition of record labels, has given these conglomerates immediate cash for paying dividends to their stockholders and has provided collateral for adding new debt. Profits in music have offset losses on movies and other properties. As one observer stated, "In contrast with $100 million film budgets, you can record Aerosmith's next album for under $1 million . . . put it on a compact disc that costs 20 cents to make, ship five million copies at $13.99, pay the band $1 per copy and draw a steady stream of cash."[42] For entertainment conglomerates that sell music, books, and movies, music sales can reinforce growth trends in movies and offset losses in books. Of course, in down years for the entertainment conglomerates, the music business can offset gains in other units. For example, within Vivendi Universal, from year-end 2001 to year-end 2002, music-division revenues decreased 4 percent, while entertainment revenues increased 27 percent, gaming revenues 21 percent, and wireless revenues 11 percent.[43]

The consolidation of contingently related businesses through mergers and acquisitions has been called "tight diversification";[44] the popular press usually refers to this as "synergy." We adopt the former term to describe the web of equity partnerships, co-brands, and joint ventures in which each of the Big Four music companies participates. Meyers succinctly describes tight diversification thus:

> A tightly diversified entertainment conglomerate is one that includes horizontally and vertically integrated firms from a range of media, including film, television, music, and publishing. These firms may include film studios, major recording companies, broadcast networks, theme parks, and Internet interests. Thus, while these firms are diversified across media and distribution platforms, they are tightly focused on entertainment.[45]

The "New Hollywood" described by Schatz makes not films, but franchises, which can be interwoven in film, music, toys, theme parks, clothing, collectables, food, and on and on.[46] Although Disney is not a Big Four music company, the *Lion King* franchise may serve as an example of tight integration working successfully between music and movie products. The *Lion King* movie earned $454 million globally; Hollywood Records, a Disney subsidiary, pressed the *Lion King* soundtrack and sold 11 million copies.[47] The New Hollywood franchise continued to roll along: "The character Simba the Lion and others were cross-licensed by the consumer products division to merchandisers who subsequently created Simba the Lion stuffed animals, tote bags, T-shirts, and . . . the Broadway musical."[48] As to the music, it may eventually be licensed to video games and ring tones. Although the cost of its music went into the *Lion King* soundtrack's first copy costs, Disney saved or recuperated a portion of these costs by keeping the licensing in-house.

Thus, unlike horizontal integration, in which a company buys its competitors, or vertical integration, in which a company buys its suppliers and distributors, tight diversification involves mergers with companies that can extend brands across different media and exhibition windows, that can secure distribution outlets, or that can provide access to new content and technologies. In tight diversification, the interlocking structures and activities of companies radiate through multiple media, changing constantly. The Sony-BMG and Vivendi-Universal mergers allowed the conglomerates involved to promote their franchises across different but related media and telecommunication companies. They were able to match content with distribution between subsidiaries, to recycle content in new windows, to extend software royalty payments through upgrades, and to implement changes in audiovisual formats they owned.

The results, such as mobile streaming audio and video (advertised as "cellevision" or "television on your cell phone") further expanded opportunities for tight diversification while at the same time allowing distribution of in-house content on in-house formats. Sony sets many of the standards for media formats. Its hardware division profits from producing equipment that creates, processes, encodes, and securely distributes digital audiovisual media under a proprietary encoding format (ATRAC3). Its music division, Sony BMG, is therefore integrated into business strategies involving consumer electronics, computer software, film, and publishing. Even though these divisions may have relative autonomy, they can be pulled together by the parent company for franchising purposes. For example,

Columbia TriStar Pictures' *Spider-Man* movie soundtrack was sold on Columbia Records (a division of Sony BMG) and generated the hit singles "Hero" by Chad Kroeger and "What We're All About" by Sum 41. Columbia licensed the *Spider-Man* character to Activision for the *Spider-Man* video game, which runs on PCs, Sony PlayStation, GameCube, and Game Boy.

Through mergers and acquisitions, record companies diversify both their products and their markets.[49] Diversification of markets, however, is more important than that of products, since spreading a smaller selection of products across a greater number of international markets can substantially improve profits. Although the United States comprises a large (and increasing) share of the global music market, U.S.-based firms such as Time Warner now operate in more than sixty countries. Multinationals Vivendi Universal and Bertelsmann have acquired U.S.-based media companies and either divested their industrial assets or packaged them into separate publicly traded companies. While market diversity increases profits, product diversity balances risks. Comparison of the Big Four's global revenues by media category suggests that they use product diversity for this purpose (see table 2.2).

To date, however, tight diversification has led to disaster as often as to success. Vivendi Universal was cobbled together from France's Vivendi (a water utility company), the Canal+ (Canal Plus) television network, and Canada's liquor giant Seagrams, which owned Universal Studios, in a "giant three-way merger" in 2000.[50] The conglomerate was near collapse by 2002 because of the debts incurred in its acquisitions, and so the company reversed its policy and began selling major assets, in the process earning the title of "France's Enron."[51] The AOL-Time Warner merger was even more disastrous. Held up, particularly in its planning stages, as the ultimate expression of tight diversification, the merger grafted an Internet-service company with

Table 2.2. Conglomerates' Content Revenues by Type, 2001 (in percentages)

	Music	*Film*	*Publishing*
Sony	32	31	—
Universal	29	16	14
Time Warner	11	21	13
Bertelsmann	19	5	28

Source: Chan-Olmstead and Chang, "Diversification," 222.

a speculative business plan onto an old company providing a vast array of media content. In addition to AOL, Time Warner owns 155 other enterprises in print, film, cable, merchandising, Internet, theme parks, and sports teams. Indeed, AOL Time Warner hailed themselves as the very harbinger of the Celestial Jukebox:

> Together, our two companies will hasten and enhance the broadband future, making real and immediate the promise of ready access to next-generation multimedia content and powerful e-commerce applications. As a result, consumers will have increased options for high-speed broadband Internet access and new means to receive new forms of content.[52]

The new company now had the exclusive right to deliver personalized performances of an enormous catalog of copyrighted culture, including such ubiquities as the "Happy Birthday" song, to online customers worldwide. Although customer adoption of broadband Internet access accelerated rapidly in portions of the United States where the infrastructure was available via Time Warner cable, the new services did not prove sufficiently popular to generate a profit. Also, as the *Wall Street Journal* revealed, a clash of corporate cultures impeded the organization's convergence strategy: "To Time Warner executives producing music, the Web makes stealing pirated copies of their products far too easy. AOL, on the other hand, has grown up in a Web culture that favors the free dissemination of everything from music to movies."[53] In pursuit of its vision of digital convergence, the AOL-Time Warner merger created an enormous debt load, which led to operating losses and a drag on the entire company. In 2003, in a desperate scramble to raise funds, AOL Time Warner sold Warner Music Group to an investment group led by Edgar Bronfman, the former CEO of Vivendi Universal.

The Warner Music sale also demonstrates that music is not as tightly integrated today into the transnational entertainment industry as it was thought to be during the 2000 megamerger of parent companies. EMI (UK) and Warner (U.S.), the little giants among the Big Four, presently are not tightly integrated with industries outside of music publishing, recording, promotion, and distribution. Although the two companies discussed a merger in 2004, Warner decided instead to prepare an initial public offering of stock shares for 2005. Still, two of the Big Four remain parts of tightly diversified conglomerates, and the other two may become so again. The potential remains for concentrating control of culture in ever fewer hands. Moreover, whatever their differences of firm-level strategies for pursuing tight integration, the major record companies continue to pursue a collective policy of locking up as much of the market as they can; they are increasingly interconnected,[54] and, as chapter 3 indicates, they now pursue a common strategy against those who challenge their control over copyrighted material—an ironic stance for supposedly competitive firms.

REPERCUSSIONS AND IMPLICATIONS: CARTEL PRICING, REGULATORY FAILURES

One of the dangers of a highly concentrated market is that its participants can easily collude with each other so as to reduce risk and increase profits while providing consumers few or no alternatives on the market, and while locking out would-be competitors with high barriers to entry. The record industry has shown itself to be much given to such collusion. Between 1991 and 2001, the Big Five raised the average price of a CD by 12.53 percent.[55] Electronics superstore chains like Best Buy and Circuit City responded by heavily discounting CDs and using them as loss leaders to entice consumers. Retailers who sold only CDs could not respond in kind. In theory, the Big Five's "minimum-advertised-price" (MAP) strategy was intended to aid those retailers by withholding cash payments intended for cooperative advertising from superstores that advertised CDs below the suggested minimum advertised price. In reality, the MAP strategy allowed the Big Five to avoid lowering wholesale costs, thus forcing retailers to maintain or increase profit margins. In May 2000, the U.S. Federal Trade Commission (FTC) ruled that the MAP strategy illegally discouraged discount pricing and inflated CD prices. It estimated that the MAP strategy cost consumers $480 million between 1996 and 1999.[56] The FTC's investigation of labels owned by Vivendi Universal led to a 2000 consent decree lasting until 2007, "wherein they agreed not to make the receipt of any cooperative advertising funds for their pre-recorded music product contingent on the price or price level at which such product is advertised or promoted."[57]

On August 8, 2000, a coalition of thirty states and U.S. territories filed suit against the record industry for price fixing.[58] In February

2002, the Big Five and three national retailers settled the lawsuit: although the companies did not admit to any wrongdoing, they agreed to refund $67.4 million to those who bought CDs from 1995 to 2000 and to donate 5.5 million CDs to nonprofit groups, schools, and libraries.[59] The payments averaged nearly $13 a piece to the more than 3.5 million people who filed claims.[60] The European Commission dropped an investigation into price fixing in the European Union in August 2001, leaving subsequent investigations to individual national authorities. They did find that three of the major companies (which they declined to name) were including minimum advertised prices in cooperative advertising with retailers in Germany, but the companies subsequently ended the practice.[61]

The Regulators Roll Over

Growing market concentration in the music industry has thus predisposed the Big Four to collusion. However, U.S. and European regulators have been inconsistent and increasingly narrow in assessing this collusion. In the minimum-advertised-price cases, for example, the European Commission tossed the issue back to national regulators. Moreover, both U.S. and European regulators have consistently approved mergers and acquisitions within the industry, despite their potential for anticompetitive behavior. Although European regulators have acknowledged the dangers of "collective dominance," in which a major record company's collection of subsidiary labels discourages competition by allowing the company to standardize CD prices, they nonetheless permitted consolidation. Similarly, the U.S. Trade Commission approved the Sony Music–BMG merger, with the sole stipulation being that the deal exclude the parent companies' music publishing, manufacturing, and physical distribution businesses and Sony's recorded-music business in Japan, Sony Music Entertainment Japan (SMEJ).[62] Sony BMG is now the world's second largest record company.

Despite the Big Five's history of collusion, U.S. Federal Trade Commissioner Mozelle W. Thompson found no reason to prevent the Sony-BMG merger. She did, however, note her misgivings:

> Although I concur in this determination, my decision was a difficult one, in part because I am particularly concerned about the impact of media mergers on the prices and quantity of media, as well as the diversity of content, available to consumers. The history of facilitating

> practices in the music industry, coupled with the elimination of Sony and BMG as independent competitors, causes me concern. . . . The industry is highly concentrated among record labels, and the proposed joint venture will only enhance this concentration. Additionally, the history of parallel MAP policies in particular indicates a propensity for interdependent behavior among the major labels.[63]

The European Commission also investigated the Sony-BMG merger for potential anticompetitive effects but did not publish the details of their findings. The commission claimed to have evidence that the major labels kept their CD prices closely aligned, even in the face of falling sales, which might have been revived by discount pricing. Although the majors published their dealer prices, the commission found that their small number and consequent market dominance gave them significant power over retailers, especially through the use of discounts to promote certain artists and products. This dominance would grow, the commission predicted, if the merger reduced the number of players from five to four.

No allegations of collusion were made directly, but the European Commission found that the Big Five used informal links and knowledge of the market to coordinate the prices they charged retailers. In an analysis of pricing structures in the European Community's largest national markets, the commission found little divergence between prices, suggesting that firms were able to affect them collectively, as would a cartel.[64] Consequently, the commission was skeptical of industry denials of collusion. The commission expressed concern that the majors would be able to extend their dominance into the fast-growing market for downloaded music, and worried about the merger's effect on concentration in the music-publishing sector, where both Sony and BMG have significant assets. The European independent record companies association, Impala, welcomed the commission's "strong objections to the merger" and praised Brussels for condemning the "anti-competitive state of the recorded music market."[65]

Nonetheless, the European Commission approved the Sony-BMG merger:

> On balance, however, the Commission had to conclude, taking into account a deficit in the transparency of the market, that the evidence found was not sufficient to demonstrate in a successful way that coordinated pricing behavior existed in the past and that a reduction from

> five to four major recording companies would not yet create a collectively held dominant position in the national markets for recorded music in the future.[66]

The commission also declined to consider online licensing and distribution, concluding that it was unwarranted in "the absence of serious competition problems": "The same goes for the examination of the vertical relationships between Sony BMG's recorded music and Bertelsmann's downstream-TV and radio activities in Germany, France, Belgium, Luxembourg and the Netherlands."[67] Affiliations of independent record labels such as Impala and labels such as Disney Corporation argued against the proposed merger and then watched and waited for antitrust action that never came. Since then, executives may have interpreted regulatory inaction as a signal that the state will acquiesce to further consolidation as a noninterventionist remedy for an "ailing" industry beset by recession, declining sales, and fluctuating stock values. Industry executives successfully invoked this "ailing-industry" rationale when the U.S. Department of Justice (also influenced by the majors' divestiture from these services) dropped an investigation of collusion between MusicNet and Pressplay; that investigation focused on the licensing practices of their parent companies to control online distribution by denying competitors access to their recordings.[68] Regulatory approvals of ongoing consolidation, despite evidence of the effects of anticompetitive practices, show that state action so far has been ineffective in protecting consumers from abuse. This regulatory failure creates new incentives for the recording industry to benefit at the expense of consumers and the public interest.

The Cultural Costs of Anticompetitiveness

Frith notes that, "because rights regimes depend on legal regulation (rather than on market forces), the economics of rights cannot be discussed separately from the politics of rights."[69] The growing consolidation within the recording industry, and the collusive behavior it has engendered, has had a chilling effect on culture. Artists suffer from the ever-expanding nature of record-company contracts, which limit releases (frequently against the artists' wishes) while exploiting vagaries in legislation to prolong artists' terms of service through exclusivity agreements. The record oligopoly's shift from goods to services via online delivery, together with regulatory inaction, introduces the possibility of another anticompetitive practice—rent seeking—to music labels offering a telecommunications-based service, which creates a new dependency between consumers (as renters) and corporations (as landlords). Rent seeking

> refers to competition for economic rents, which are returns in excess of costs. Such returns are generally available only via monopolistic market positions. . . . What is distinct about rent-seeking versus the more typical sort of economic competitiveness is that rivals vie not by producing new or preferred outputs for consumers, but by influencing authorities to divert a fixed resource to their ownership. Hence rent-seeking is often identified as classically wasteful, in that no net addition to social well-being develops from the competitive struggle.[70]

An example of rent seeking may be found in the cable-television industry. Franchises for cable service to specific areas are awarded by city councils to a single company. If this company subsequently engages in rent seeking, it will exploit its status as sole service provider by raising prices, cutting services, or failing to respond to customer demands for diverse and innovative services or content. While cable providers operate as a monopoly, the Big Four operate as an oligopoly; nevertheless, by acting in concert in setting rates for licensing recordings to online providers, or in withholding recordings, they hold the potential for rent seeking with the Jukebox model for media distribution. Public-interest regulation has (in the past) largely constrained the cable industry from the grossest excesses of consumer abuse; it should likewise control the recording industry if technology gives it similar opportunities for monopolistic exploitation of musical culture.

The recording industry emerged in the early twentieth century from collusion over patents;[71] today, its practices are increasingly based in collusion over copyrights. The largest recording companies are now parts of conglomerates that—although involved in diverse media including film, television, music, and publishing—are focused on entertainment and copyright. Digital technology has made risk management increasingly problematic for copyright holders; diversity within and competition among these conglomerates have so far interfered with the harmonization of technical standards and company policies. However, these conglomerates have united behind legal and legislative efforts to protect their intellectual-property assets, and we can expect greater harmony among them in formally "competitive" areas, such as hardware and software, as the industry value chain becomes aligned.

Contrary to marketing and public-relations hyperbole, the Celestial Jukebox does not bring us "revolutionary" technology and culture. Instead, as we show in chapter 3, it has extended the music-industry oligopoly into cyberspace, in the process suppressing the music scenes, which arise from communities working together, it has found there. In 1995, Prince predicted that "once the Internet is a reality, the music business is finished. There won't be any need for record companies. If I can send you my music direct, what's the point of having a music business?"[72] Prince's prediction for the demise of record companies is as likely to be realized as his prediction of Armageddon in "1999"; it seems unlikely that the Celestial Jukebox will change any structural relationships between label and artist, which is "one of the most grotesquely polarized in the world."[73] Nor will it lead to an abundance of cultural pluralism and diversity. Many consumers will add another utility bill (from a music-service provider or clearinghouse) to their monthly budget, and record labels will still be able "to extract modest amounts of money from vast numbers of people"[74] in a rent-seeking activity tacitly approved by the state.

In chapter 3, we consider the exit options available to citizens in an information society, including a turn to the Darknet for copyrighted culture. We also investigate the Big Four's production of cultural scarcity in cyberspace in more detail. The centralizing systems of computer and networking technology, closed-loop production processes, and proprietary formats and platforms constitute a cumulative loss of ground to the narrow interests of a highly concentrated entertainment industry. The shift from hard media to soft media could have made fuller use of the open architecture and shared intelligence the Internet offers. Instead, it has inflicted economic, social, and cultural harm.

NOTES

1. Harold L. Vogel, *Entertainment Industry Economics*, 4th ed. (Cambridge, MA: Cambridge University Press, 1998).

2. Recording Industry Association of America, "RIAA Issues 2004 Year-End Shipment Numbers," *Press Room*, 21 Mar. 2005, http://www.riaa.com/news/newsletter/032105.asp.

3. Philip H. Ennis, *The Seventh Stream: The Emergence of Rock'n'Roll in American Popular Music* (Hanover, NH: Wesleyan University Press, 1992); Russell Sanjek, *American Popular Music and Its Business*, vol. 3, *From 1900 to 1984* (New York: Oxford University Press, 1988); Russell Sanjek and David Sanjek, *American Popular Music Business in the 20th Century* (New York: Oxford University Press, 1991).

4. Sanjek and Sanjek, *American Popular Music*.

5. Eric Rothenbuhler and Tom McCourt, "The Economics of the Recording Industry," in *Media Economics: Theory and Practice*, 3rd ed., ed. Alison Alexander et al. (Mahwah, NJ: Lawrence Erlbaum, 2004).

6. Doyle, Gillian, *Understanding Media Economics* (Thousand Oaks, CA: Sage Publications, 2002).

7. Eric Leach and Bill Henslee, "Follow the Money: Who's Really Making the Dough?" *Electronic Musician*, 1 Nov. 2001, http://industryclick.com/magazinearticle.asp?magazineid=33&releaseid=9554&magazinearticleid=132835&SiteID=15.

8. M. William Krasilovsky and Sidney Shemel, *This Business of Music: The Definitive Guide to the Music Industry*, 8th ed. (New York: Billboard Books, 2000).

9. Krasilovsky and Shemel, *This Business*, 14.

10. Chuck Philips, "Lawmakers Take Aim at Music Industry Contracts," *Los Angeles Times*, 8 Aug. 2001, C1. http://latimes.com/business/la-000064238aug08.story?coll=la%2Dheadlines%2Dbu.

11. Philips, "Lawmakers."

12. Jennifer Ordonez, "Musicians, Record Companies Face Off," *Wall Street Journal*, 5 Sept. 2001, B5.

13. Corey Moss, "Beck, Deftones, Others Rally for Bill that Could Change Recording Contracts," *Indie-music.com*, 2 Feb. 2002, http://www.indie-music.com/modules.php?name=News&file=article&sid=819.

14. Courtney Love, "Courtney Love Does the Math," *Salon*, 14 June 2000, http://tinyurl.com/2x25e; Steve Albini, "The Problem with Music," *NegativWorldWideWebland*, http:www.negativland.com/albini.html.

15. Neil Strauss, "Behind the Grammys, Revolt in the Industry," *New York Times*, 24 Feb. 2002, WK3.

16. Chuck Philips, "Dixie Chicks, Sony End Feud with a New Deal," *Los Angeles Times*, 17 June 2002, C1.

17. Philips, "Lawmakers."

18. Charles C. Mann, "The Heavenly Jukebox," *Atlantic Monthly*, Sept. 2000, 52.

19. Mann, "Heavenly Jukebox," 50.

20. Vogel, H, *Entertainment Industry Economics*, 4th ed. (New York: Cambridge University Press, 1998).
21. Leach and Henslee, "Follow the Money."
22. Mann, "Heavenly Jukebox," 50.
23. Rothenbuhler and McCourt, "Economics."
24. Reebee Garofalo, "From Music Publishing to MP3: Music and Industry in the Twentieth Century," *American Music* 17, no. 3 (Fall 1999): 321.
25. Garofalo, "From Music Publishing," 322.
26. Adam David Marcus, "The Celestial Jukebox Revisited: Best Practices and Copyright Law Revisions for Subscription-Based Online Music Services" (unpublished paper, 2003), 8, http://tprc.org/papers/2003/220/TPRC_paper-Adam_Marcus.pdf.
27. Garofalo, "From Music Publishing," 327.
28. Garofalo, "From Music Publishing," 328.
29. Marcus, "Celestial Jukebox," 9.
30. Marcus, "Celestial Jukebox," 9.
31. Paula Samuelson, "The Copyright Grab," *Wired*, http://www.wired.com/wired/archive/4.01/white.paper_pr.html.
32. Simon Frith, "Video Pop: Picking up the Pieces," in *Facing the Music*, ed. Simon Frith (New York: Pantheon, 1988), 88–130.
33. Garofalo, "From Music Publishing," 319 (bracketed dates added by author).
34. Andre Millard, *America on Record: A History of Recorded Sound* (New York: Cambridge University Press, 1996), 65.
35. Ben Bagdikian, *The Media Monopoly*, 6th ed. (New York: Beacon Press, 2000).
36. Lee Isgur, in Bagdikian, *The Media Monopoly*, 4.
37. Paul D. Lopes, "Innovation and Diversity in the Popular Music Industry, 1969 to 1990," *American Sociological Review* 57 (1992): 56–71; Robert Burnett, "The Implications of Ownership Changes on Concentration and Diversity in the Phonogram Industry," *Communication Research* 19 (1992): 749–69.
38. Richard A. Peterson and David G. Berger, "Cycles in Symbol Production: The Case of Popular Music," *American Sociological Review* 40 (1975): 158–73.
39. C. Goldsmith, M. Karnitschnig, M. Peers, and B. Orwall, "As Music Sector's Woes Worsen, Sony and BMG Propose a Merger," *Wall Street Journal*, 7 Nov. 2003, A1.
40. Peterson and Berger, "Cycles"; Eric W. Rothenbuhler and John Dimmick, "Popular Music: Concentration and Diversity in the Industry, 1974–1980," *Journal of Communication* 32, no. 1 (1982): 143–49; and Lopes, "Innovation."
41. Andrew Leyshon, David Matless, and George Revill, eds., *The Place of Music* (New York: The Guilford Press, 1998).

42. Andy Kessler, "MP3 Breaks Records. Look Out, AOL," *Wall Street Journal*, 25 Jan. 2005, A18.

43. Vivendi Universal, "Vivendi Universal Preliminary Supplemental Revenues Information," 2003, http://finance.vivendiuniversal.com/finance/documents/financial/annual-release/2002/PR100203Charts.pdf.

44. Tom Schatz, "The Return of the Hollywood Studio System," in *Conglomerates and the Media*, ed. Patricia Aufderheide et al. (New York: W. W. Norton, 1997).

45. Cynthia Meyers, "Entertainment Industry Integration Strategies" (paper presented at the Institute for International Studies and Training, Tokyo, Japan, 14 Feb. 2002), http://www.iist.or.jp/wf/magazine/0068/0068_E.html.

46. Schatz, "The Return."

47. Richard A. Gershon, "The Transnational Media Corporation: Environmental Scanning and Strategy Formulation," *The Journal of Media Economics* 13, no. 2 (2000): 94.

48. Gershon, "Transnational," 95.

49. Sylvia Chan-Olmstead and Byeng-Hee Chang, "Diversification Strategy of Global Media Conglomerates: Examining Its Patterns and Determinants," *Journal of Media Economics* 16, no. 4 (2003): 213–33.

50. John Carreyrou and Martin Peers, "Damage Control: How Messier Kept Cash Crisis at Vivendi Hidden for Months," *Wall Street Journal*, 31 Oct. 2002, A1.

51. Carreyou and Peers, "Damage Control," A1.

52. Quoted in Patricia Aufderheide, "Competition and Commons: The Public Interest in and after the AOL-Time Warner Merger," *Journal of Broadcasting and Electronic Media* 46, no. 4 (2002): 522.

53. Martin Peers and Nick Wingfield, "Seeking Harmony, AOL and Warner Music Hit Some Dissonant Notes," *Wall Street Journal*, 18 Apr. 2000, B1.

54. Chan-Olmstead and Chang, "Diversification," 214.

55. John Snyder, "Embrace File-Sharing, or Die," *Salon*, 1 Feb. 2003, http://www.salon.com/tech/feature/2003/02/01/file_trading_manifesto/print.html.

56. John Wilke, "Music Firms, U.S. Hold Settlement Talks," *Wall Street Journal*, 16 Dec. 1999, A3; Martin Peers and Evan Ramstad, "Price of CDs Likely to Drop, Thanks to FTC," *Wall Street Journal*, 11 May 2000, B1.

57. Vivendi Universal, Form 20-F of Annual Report Pursuant to Section 13 or 15(d) of the Securities Exchange Act of 1934 (2004), 31.

58. Martin Peers, "States Sue Record Companies over CD Pricing Policies," *Wall Street Journal*, 9 Aug. 2000, B7.

59. Claudio Deutsch, "Suit Settled Over Pricing of Music CDs at 3 Chains." *New York Times*, 1 Feb. 2002, C1.

60. CNN.com, "Judge: Millions of CD Buyers Owed Money," 16 June 2003, reprinted at http://www.techzonez.com/forums/showthread.php?t=5478.

61. *New York Times*, "European Inquiry into CD Price Fixing Ends," 18 Aug. 2001, C14.

62. *Reuters*, "Sony Music, BMG Complete Merger," 5 Aug. 2004, http://news.yahoo.com/news?tmpl=story&u=/nm/20040806/bs_nm/media_sony_bmg_dc_3.

63. Mozelle W. Thompson, "Statement of Commissioner Mozelle W. Thompson," Sony Corporation of America/Bertelsmann Music (Group Joint Venture File No. 041-0054, 2004), http://www.ftc.gov/os/caselist/0410054/040728mwtstmnt0410054.pdf, 1.

64. Simon Taylor, "Media Business: Brussels Says Sony-BMG Pact Would Worsen Flawed Market: EC Report Claims Big Five Recording Groups Already Control Retail Prices," *The Guardian* (London), 11 June 2004, 23.

65. Taylor, "Media Business," 23.

66. European Commission, "Commission Decides Not to Oppose Recorded Music JV between Sony and Bertelsmann," Press Release IP/04/959, 20 July 2004, http://europa.eu.int/rapid/pressReleasesAction.do?reference=IP/04/959&format=HTML&aged=0&language=EN&guiLanguage=en.

67. European Commission, "Commission Decides."

68. Scarlet Pruitt, "Online Music Probe May Be Nothing New," *PC-World.com*, 8 Aug. 2001, http://www.pcworld.com/news/article/0,aid,57403,00.asp; Mark Wigfield, "Probe of Web Music Ventures Ends," *Wall Street Journal*, 24 Dec. 2003, B4.

69. Simon Frith, "Music Industry Research: Where Now? Where Next? Notes from Britain," *Popular Music* 19, no. 3 (2000): 387–93, 388 (parentheses in original).

70. Thomas W. Hazlett, "Rent-Seeking in the Telco/Cable Cross-Ownership Controversy," *Telecommunications Policy* 14, no. 5 (Oct. 1990): 425–26.

71. Garofalo, "From Music Publishing."

72. Quoted in Philip Hayward, "Enterprise on the New Frontier: Music, Industry, and the Internet," *Convergence* 1, no. 2 (Autumn 1995): 32.

73. John Lovering, "The Global Music Industry Contradictions in the Commodification in the Place of Music," in *The Place of Music*, ed. Andrew Leyshon, David Matless, and George Revill (New York: The Guilford Press, 1998), 41.

74. Lovering, "Global Music Industry," 41.

3

The Jukebox Contested

When the companies that dominate an industry find their status challenged by new technologies, their enduring response is to incorporate, contain, or destroy these technologies through forces of the market and the state. New technologies, which foster the creation of alternative models of distribution and promotion, have been opposed by the Big Four because the very possibility of these models threatens their primary strengths. As noted in chapter 2, the Big Four's cartel relations and market "parallelisms" (in the terminology of the European Commission) have allowed them to bring in "substantial flows of income" from activities that exercise "no creative function."[1] Having concentrated within themselves tremendous market power, political power, and cultural power, the Big Four are now in the process of securing, through the Celestial Jukebox, an equally concentrated "network power."[2]

Chapter 2 described the structure of the recording industry, a structure that the Big Four have found immensely profitable and plan to extend into cyberspace. In this chapter, we examine the "disruptive" technologies that challenge that plan, and with it the Big Four's hegemony in the recording industry. A "disruptive technology"—as we use the term—upsets a long-standing business model in an existing market. In the process, it alters established relationships between industry stakeholders[3] and threatens to superannuate or supplant markets under their control. File sharing fundamentally challenges the Big Four by threatening to render their greatest strengths, distribution and promotion, obsolete. Their responses to file sharing reveal a great deal about what social goods are at risk as the industry moves online. To succeed, the Big Four's Celestial Jukebox must overcome the Internet's open and decentralized architecture, as well as the social and community-oriented groups that seek to maximize and defend civil liberties in cyberspace. What it cannot control it must co-opt; Napster and its cousins shaped the contours of the Celestial Jukebox's design. Ironically, at least in the short term, the record industry's attempts at containment have actually enhanced the usability, effectiveness, and status of these disruptive technologies, thereby diminishing chances for the Jukebox's adoption.

DISRUPTIVE TECHNOLOGIES

New wired and wireless communications technologies have always brought about structural changes in commerce. The telegraphy system resembled a "Victorian Internet,"[4] in which telegraph offices received, copied, and forwarded messages until they reached the recipient's locale. Telegraphy disrupted common carrier service for the mails and optical signaling for military communications. The telephone's direct circuit switching to customers in turn disrupted and ultimately absorbed the telegraph. Similarly, Internet packet switching disrupted telephony, both through the new format of e-mail and the voice-over-Internet transmission that bypassed the telephone network. An exemplar of this process was AT&T, which after 1996 was transformed from a long-distance carrier to a networking company, and then, in 2005, was taken over by SBC, which was once a "Baby Bell" spin-off from the AT&T divestiture.

This ability of new technology to challenge dominant power structures has in recent years brought about a kind of Internet nirvana theory. The many celebrants of Napster believed that technology alone would usher in an era of "post-scarcity" and provide an unlimited array of goods and services. These predictions of a "cornucopia of the commons [in which] use brings overflowing abundance"[5] have not been realized, yet cyberutopianism continues, for example, in speculation about wireless ad hoc networks. As WiFi and Bluetooth grow in popularity, they are predicted to defy the tragedy of the commons, in which individual aims and actions exhaust a collective resource. The utopian vision instead gives us visions of "sheep that shit grass."[6]

Yet it is inarguable that new media technologies have disrupted the recording industry by widening or eliminating the gateways that that industry once exclusively, and profitably, maintained. First, they have allowed recordings to bypass traditional distribution channels and thereby evade the industry's control over access to its intellectual property. Second, they have provided delivery platforms the industry does not control. Though they may not bring about a future of utopian happiness, peer-to-peer systems, and the social and political interests they represent, continue to undermine the regime that seeks to consolidate network power around the Celestial Jukebox. Despite legal setbacks, citizens and consumers will continue to question the value of suppressing Internet distribution technologies for the benefit of industry.

The MP3 Standard

WAV files were the earliest format of digital-content storage suitable for delivery by modem and PC; however, three-minute songs in this format took hours to download. In 1987, the Motion Picture Experts Group (MPEG), of the Geneva-based International Organization for Standardization, began to develop ways to compress digital video, shrinking the signal to ease transmission. The MPEG framework enabled the Fraunhofer Institute, a German company working in digital television, to develop a data-compression algorithm for audio that delivered a high-quality sound recording at a very low bit rate or encoding quality. The resulting format, MP3, was released in 1992. MP3 files are denser than raw audio data, which they compress from standard CD (Red Book) quality by a factor ranging from five-to-one to twenty-two-to-one, depending on the bit rate. MP3 makes it possible for the average Internet user to participate quite easily in the exchange of audio files, particularly as modem speeds increase. Using variable bit rates, a 56K telephone modem can transfer one four-minute song in about eleven minutes, and a cable modem can do it in about forty-eight seconds; faster links make transfer time almost a nonissue.[7]

Cheap and easy to encode, decode, and distribute, the MP3 format helped enable file sharing of music. Although the format is too "lossy" to be useful to artists in production, its lower quality is an acceptable trade-off for convenience in file sharing. Significantly, the MP3 format offers no intrinsic protection against copying; MP3s "ripped" from commercially released CDs are easily exchanged on the Darknet (described below). MP3s on the Web therefore disrupted the music industry by holding out the "possibility of a business model that links artists directly to consumers, bypassing the record companies completely."[8] One of the earliest venues for such activity, Internet relay chat (IRC), allowed semianonymous users to trade MP3s directly. Such trading quickly demonstrated the social value of exchanges in cultural capital on the unstructured Internet. In this "audio-piracy subculture,"[9] members gained elite status by providing large numbers of MP3 files to an "exclusionary and status driven" virtual community.[10]

As the supply of MP3s increased and broadband access grew more common, large numbers of music fans joined virtual communities for online sharing. At the close of the 1990s, the popular press ran a spate of stories about MP3 as a (primarily) youth-culture phenomenon likely to lead to trouble. The *Wall Street Journal* reported in March 1999 that "MP3 has created an underground online culture, in which hackers hang around chat rooms and online 'gangs' prowl for tunes."[11] Another article quotes a user: "'At my school, almost everyone who has an Internet connection has MP3s,' says Brendan, who himself has 1½ gigabytes of MP3 music, the equivalent of several hundred singles."[12] The recording industry reacted at first as Hollywood had to Betamax and other technologies outside of their immediate control: it sued. In late 1998, Diamond Multimedia introduced the Rio, a Walkman-sized portable MP3 player retailing for two hundred dollars, which attached to a computer and downloaded up to sixty minutes of sound files. The RIAA sued for an injunction against Diamond on the grounds that the player violated the Audio Home Recording Act (AHRA) of 1992. A federal judge, however, ruled against the RIAA on October 26, 1998, and Diamond Multimedia cranked up its production.[13]

MP3's present dominance is threatened both by those who seek to open and those who seek to restrict free access to music. MP3 itself is not free. The Fraunhofer Institute charges licensing fees to anyone who creates MP3 hardware or software or sells MP3 downloads over the Internet; hardware companies pay fifty cents per unit shipped, while download companies pay 1 percent of royalties. Ogg Vorbis, an open-source project, has sought to create a free replacement for MP3 while technically exceeding that format's specifications. The recording industry, meanwhile, has sought to replace MP3 with new proprietary systems (such as Sony ATRAC, Apple FairPlay AAC, RealAudio, and Windows Media Audio) that afford greater protection against copying. So far, they have met with little success, as MP3 is more versatile and functional than these corporately owned technologies. While MP3's momentum has hampered the establishment of such copy-protected successors, mainstream producers increasingly use them. Of the sanctioned online music-service providers, only eMusic distributes their files in the MP3 format; in the future, MP3 encoded music may become relegated to the Darknet.

Peer-to-Peer and the "Gift Economy"

Like MP3, peer-to-peer technologies open a variety of gateways once controlled by information industries. In peer-to-peer networking, each computer operates as both a client and a server for other computers in its network. The network is informal; it does not require any single computer to coordinate traffic, and it shrinks and grows as computers join or drop off. Peer-to-peer is not new; its networking model formed the very foundation of the Internet. As Minar and Hedlund note,[14] Usenet newsgroups implemented "a decentralized model of control that in some ways is the grandfather of today's new peer-to-peer applications such as Gnutella and Freenet."

The World Wide Web also uses P2P networking technology. The Internet Domain Name System (DNS) makes it possible for people to enter user-friendly domain names into their browsers so that they do not have to enter impossible-to-remember Internet Protocol (IP) addresses. DNS makes it possible to enter a Uniform Resource Locator (URL) such as "iTunes.com" or "Stanford.edu" into a Web browser, rather than a long combination of numbers such as "165.91.22.81." DNS intercepts these user-friendly URLs and converts them into network addresses that the Internet understands. In so doing, it combines peer-to-peer networking with "a hierarchical model of information ownership."[15] Unlike e-mail and other Internet technologies that depend on servers, peer-to-peer networking decentralizes features, infrastructure, and administration. Peer-to-peer typically occurs at the "edges" of the network, where Internet addresses are generated and temporarily assigned to computers acting as nodes. Peer-to-peer nodes can also have fixed IP addresses and function on networks as "supernodes." Given the constantly mutating nature of peer-to-peer networks, there is no fixed function for any given user, and DNS does not provide a naming scheme for peer-to-peer locations.[16]

The DNS architecture that matches domain names with IP addresses introduces a basic level of scarcity in cyberspace; it is therefore responsible for the political economy of the Internet.[17] Peer-to-peer technology, however, disrupts the fixed-IP universe of DNS addressing. Its addressing bypasses DNS, shifting control of file transfers away from centralized servers and to the nodes themselves. When this happens, as with instant messaging (the low bandwidth requirements of which made it one of the first peer-to-peer technologies), "your address has nothing to do with the DNS hierarchy, or even with a particular machine, except temporarily; your chat address travels with you."[18] In its nascent stages, music distribution on the Internet among Internet start-ups, fan sites, and band sites was like P2P networking—an informal, experimental, and idealistic affair.[19] Some of the earliest music outposts, such as the Communal Groove Machine, Nettwerk, Indie Oasis, and Internet Underground Music Archive (IUMA), are now nearly forgotten. Drawing on the "sneaker-net" of home tapers who distributed tapes or disks among friends, file traders developed social practices based on a collective belief that the doctrine of fair use and noncommercial distribution of media content extended to the Internet. The doctrine was enshrined in the 1992 Audio Home Recording Act, section 1008, which "makes explicit that consumers do not infringe by making noncommercial digital or analog musical recordings"[20]:

> No action may be brought under [the Copyright Act] alleging infringement of copyright based on the manufacture, importation, or distribution of a digital audio recording device, a digital audio recording medium, an analog recording device, or an analog recording medium, *or based on the noncommercial use by a consumer of such a device or medium for making digital musical recordings or analog music recordings.*[21]

As multimedia PCs with MP3 encoders and Internet access became more prevalent, peer-to-peer networking technology created the basic exchange relationship in cyberspace,[22] and the recording industry began to feel ever more threatened. Since recordings already existed in digital form (compact discs), they were easily converted to MP3s for peer-to-peer distribution. The ensuing "gift economy"[23] challenged the economics of the recording industry. The new communication technologies appeared to offer radical opportunities for entrepreneurs, innovators, ideologues, and upstarts to transform the industry, much as telephony had transformed telegraphy, audio recording had transformed the sheet-music industry, and radio had transformed the recording industry. Peer-to-peer users believed, not without some legal basis in recent history, that they had the "right to make home copies of recorded music for personal use."[24] However, as a public-distribution platform, peer-to-peer systems challenged the recording industry's exclusive control over the distribution of sound recordings and the "musical works embodied in the files."[25]

The sharing communities enabled by P2P networks are economically, culturally, and technologically incompatible with the online music stores through which the recording industry hopes to control access to their properties. Copies of digital media proliferate freely in network environments because sharing music is a basic ritual held in common across cultures, and because making a digital copy of a recording is a simple operation that can be accomplished by an inexperienced computer user. Peer-to-peer systems created an alternative Celestial Jukebox, featuring a substantial catalog of content that is freely available and accessible at all times, before the Big Four constructed a commercial version. The Big Four reacted by declaring these systems, which are based on the sharing principle of the Internet commons, antisocial and illegal.

In consequence, these seemingly innocuous noncommercial gift exchanges—and their enabling technologies—are now among the scariest monsters in all of cyberspace, outlawed, or nearly so, because they can jump the fence that media oligopolists have constructed around intellectual property. Years after Napster's shutdown, PR agencies and legal counsel for the entertainment industry still repeatedly claim that peer-to-peer threatens its survival, just as their predecessors in the 1980s claimed that "home taping kills the music industry." The new intellectual-property regime for the digital domain, codified in the DMCA and certified in the Napster case (discussed later in this chapter), says that there may be no free gifts in cyberspace. Online music stores, barrages of lawsuits, and tighter

DRM controls will, the oligopolists hope, enforce a new regime, a "pay-per" society, in which the intellectual commons is privatized and meted out for profit by themselves.[26]

DEALING WITH DISRUPTION

Online distribution of content in "virtual" files is crucially different from the physical distribution of hard goods that has long shaped recording-industry practices. Physical distribution rests on a delicate balance between maintaining tight controls on inventory and ensuring that this inventory is adequately distributed, despite pressures to limit supplies to stores.[27] Selling CDs is similar to selling produce or retail clothing; the inventory must be managed effectively to achieve an optimum price, before the goods go bad or stale. To achieve this end,

> Staff in the distribution division of the major labels work at the "interface" between record company and retailer, and include market researchers, sales staff and business analysts. Their task is to monitor stock movements within the company's warehouses and among different retail outlets, and to ensure that the company is not pressing too many recordings (and wasting valuable storage space) or making too few recordings (and losing money by failing to respond to public demand).[28]

In cyberspace, these concerns vanish. Goods may be copied and transported over the Internet at marginal costs, and unwanted goods may easily be discarded or voided. By implementing rights-managed distribution on the Internet, the Big Four can buy, sell, and resell audiences and intellectual property in a kind of market arbitrage. This arrangement is possible because, "without the material substratum restraining them, commodities may respond instantly to the fractal climate of fashion."[29] Napster, at its peak, illustrated this "fractal climate": Andy Greenwald of *Spin* wrote, "[Napster's] very nature—the trading of one song at a time—will place an emphasis on singles. In colleges one song tends to make a hot list, sweep the campus, and then be replaced by another the next week."[30] Envisioning a plethora of online packages for consumers, Edgar Bronfman predicted, "You'll be able to program bundles or song packages, compilations, video singles and video compilations. You'll be able to buy or program songs by genre, by era, by the hour or half-hour or minute or day."[31]

Both pay-per-use and subscription models can present "live" performances like broadcasting, or "shift" the time and place at which recorded performances are consumed. Time- and space-shifting technologies are increasingly commonplace; they include videocassette recorders, portable audiocassette recorders and players, TiVo, and peer-to-peer networks. All of these technologies increase opportunities for access and consumption and have therefore been wildly popular with consumers. However, since they threaten existing industrial models and practices, these technologies are increasingly subject to a new economic system—a client-server system based on pay-per-use, which is now delivering content on demand to households and mobile devices. As cell phones, cable set-top boxes, video-game units, PCs, and other addressable devices are stretching opportunities for time and space shifting, they also create new "windowing" options for distribution.

The privatization of the Internet commons is crucial to the success of pay-per-use. The early Internet functioned as a peer-to-peer system in a manner that was

> more open and free than today's networks. Firewalls were unknown until the late 1980s. Generally, any two machines on the Internet could send packets to each other. The Net was a playground of cooperative researchers who generally did not need protection from each other.[32]

A large proportion of the Internet is now composed of private networks, and this proportion is increasing as more and more networks are migrating from egalitarian peer-to-peer to privileged client-server architectures. The openness and interconnectivity of the Internet is under pressure from private and commercial interests. The Internet figures larger and larger in the plans of these interests, and organizations and practices that fall outside of their authority increasingly face an uphill battle.

Scour and MP3.com

The cases of Scour and MP3.com illustrate the Big Four's conviction that any alternative to the Celestial Jukebox is a direct threat to their hegemony and must be suppressed and/or co-opted. Scour, a search engine for sound and video files, was created by a group of UCLA students in 1997. Two years later, Hollywood superagent Michael Ovitz and a partner bought a majority share of the service, and in April 2000, Scour added a free video and audio file-sharing service (Scour Exchange). Scour was averaging 443,000 visitors a month by May 2000 and ultimately attracted more than 7 million.[33] The RIAA, the MPAA, and the National Music Publishers Association (NMPA) sued Scour in July 2000 for copyright violations (ironically, these organizations represented companies that Ovitz dealt with in his other ventures). After the lawsuit scared off potential investors, Scour laid off fifty-two of its seventy employees in October 2000 and filed for bankruptcy, shutting down its file-sharing program a month later.[34] Scour's remains were purchased for $9 million by CenterSpan, which envisioned a new "pay" system that incorporated DRM. The new service was to be "paraded around in January [2000] to potential investors and business partners,"[35] but it ultimately came to nothing. Scour's fate indicates how startups may be acquired to attract short-term venture capital to parent firms, though long-term profits are far from certain.

Another potential alternative to the Celestial Jukebox, MP3.com, provided a distribution conduit for unsigned musicians who provided content to the company free of charge. MP3.com earned revenues from running banner ads, selling consumer data to DoubleClick and other companies, and splitting user micropayments with musicians. After an initial public stock offering that reaped $6.9 billion dollars (which was half a billion dollars higher than EMI's market valuation at the time),[36] MP3.com launched two commercial applications in January 2000: Instant Listening and Beam-It. These services offered free online cyberlockers with a capacity of twenty-five CDs (which could be increased by paying an annual fee) that could be personalized with the MyMP3.com service. The lockers did not offer downloads; instead, a user could log on from any Internet-connected computer and obtain streamed music. To access these streams, the user had to verify ownership of a CD, either by inserting it into the CD-ROM drive of a computer, or by purchasing it online from one of MP3.com's retail partners, such as Tower Records, after which MP3.com would place a copy of the music into the person's virtual locker.[37] MP3.com launched the service without seeking licenses from record companies; the major record companies objected that users could create accounts with borrowed CDs, since there was no way to verify ownership.[38]

Therefore, in January 2000, despite the company's good-faith effort to guarantee ownership and fair use of CDs, the RIAA sued MP3.com, alleging copyright infringement. (Ironically, MP3.com CEO Michael Robertson had also signed the RIAA's request for an injunction against Napster.) On April 28, a federal judge ruled that MP3.com had violated copyright law.[39] Ten days later, BMG unveiled Music Bank, which used MP3.com's storage-locker model to allow on-demand streaming access to BMG's entire music catalog. On June 10, MP3.com settled with BMG and Warner Music Group, agreeing to pay each company approximately $20 million for use of their music on MyMP3.com.[40] MP3.com also agreed to pay 1.5 cents each time a user stored a song on the service, and one-third of a cent each time it was accessed from the MP3.com website, fees that might amount to tens of millions of dollars a year. MP3.com later reached settlements with Sony and EMI for an estimated $20 million each.[41] The sole holdout, Universal, asked for $450 million in damages. On August 6, 2000, Judge Jed Rakoff ruled that MP3.com infringed upon Universal's copyrights and would have to pay $250 million, or $25,000 for each of the Universal compact discs included in MyMP3.com.[42] According to a filing with the Securities and Exchange Commission (SEC), MP3.com had about $405 million in total assets at the time. In October, the company announced that it would pay up to $30 million to music publishers, which settled a lawsuit brought by the Harry Fox Agency, the licensing subsidiary of the National Music Publishers Association.[43]

On November 14, 2000, MP3.com agreed to settle the lawsuit with Universal for $53.4 million, the largest sum the company could pay without going bankrupt.[44] According to industry observers, MP3.com was able to settle for significantly less than $250 million because a trial might have revealed damaging information about Universal's business practices, specifically that it allegedly and regularly defrauded the U.S. Copyright Office by unlawfully taking advantage of the work-for-hire clause in artist contracts. A memo by MP3.com lawyers indicated that artists and producers were producing work for hire without checking their agreements, that Universal registered recordings as "remixes" without determining if any new material was added, and that the company registered works as compilations even if it failed to own the copyrights.[45] Ironically, Universal was in turn sued by the National Music Publishers Association, representing songwriters and music publishing firms who claimed that the company was using recordings of their compositions on its

Farmclub music-streaming service without licensing them first—the very practice that led Universal to sue MP3.com.[46]

Although the Big Five thus protected their copyright interests, few licensing fees probably went to the artists in whose behalf the RIAA claimed to be acting. Universal held that Internet downloads were covered by the same license as record clubs, for which artists are not owed any proceeds. Nor, despite the RIAA's rhetoric that artists "have the fundamental right to decide which innovative business models they want to pursue and which they do not," were artists consulted either by MP3.com or the Big Five.[47] A few acted on their own. Randy Newman, Tom Waits, and Heart filed suit against MP3.com for $40.5 million in May 2001, claiming that MyMP3.com provided illegal access to their recordings. They asked for a maximum penalty of $150,000 per song.[48]

After the Universal settlement, MP3.com relaunched its storage-locker service, in both free and subscription versions. In the former, a user could store up to twenty-five CDs free but would receive text and audio ads; in the latter, users would pay $49 a year to access a catalog of CDs. The result, as the *New York Times* dryly noted, was "the joy of paying to listen to CD's you already own" in versions of inferior MP3 quality.[49] Then, in an abrupt about-face, Vivendi Universal purchased MP3.com for $372 million in May 2001, six months after the settlement. The finances behind the acquisition are instructive. MP3.com's stock was valued as high as $63.61 before a barrage of copyright-infringement lawsuits from artists, publishers, and record labels hit the company. Vivendi Universal offered $5 per share for MP3.com's stock, which had traded for just $3.01 per share before the acquisition was announced.[50]

Despite its legal liabilities, MP3.com was attractive to Vivendi Universal because it was one of the few firms with the technological infrastructure in place to operate a large-scale online distribution service, and also because it had a well-known brand and a sizable user base. Shortly after the acquisition, Vivendi Universal split the company between the MP3.com website and a new division that would provide technology for its planned Pressplay service.[51] However, cash-strapped Vivendi closed the European division of MP3.com in August 2003, and the site was acquired by CNet in November, which turned it into a music reference guide. The archive of over 1.7 million songs was turned over to Trusonic, a Vivendi spin-off, and Garageband.com.[52] The songs, representing the toil of thousands of aspiring artists, are now lost in cyberspace.

The Napster Imbroglio

MP3.com was simultaneously suppressed through litigation and co-opted through adaptation of its technology, but it never posed a serious threat to the recording industry. That threat, and the Big Five's true bête noire, was Napster. Created by Shawn Fanning, a nineteen-year-old college dropout, Napster functioned as a music search engine that linked participants to a huge and constantly updated library of MP3s provided by users. Napster's key architectural feature was an online database of song titles and performers, searchable by keyword;[53] the network's MusicShare client provided access to the indices and file lists of those using the service. Its brokered architecture effectively coordinated users and increased search functionality, and its search-and-play interface was highly user-friendly. Napster also featured a virtual community for music fans. The program was released on the Internet in August 1999, and the resulting glut of digital traffic overloaded university computer networks. In fact, the service was quickly banned from several universities because of the strain on networks.

No sooner had Napster become a "killer app" than legal woes beset the company. The RIAA filed suit against Napster on December 7, 1999, claiming that the free service cut into sales of CDs. Napster's enabling architecture became its legal vulnerability: when a computer with Napster software was connected to the Internet, it became both a receiver, or client, and a sender, or server, and its user became a publisher as well as a consumer. The legal case against Napster turned on the fact that although it did not generate revenue, it supplied users with software for sharing files and a real-time index of available music files. This combination of marketed products and services, the RIAA argued, effectively turned Napster into a music-piracy service. While providing an addressing function independent of DNS was perfectly legal, as was creating a search engine, the RIAA argued that the combination of these legal technologies created an illegal result that would destroy the recording industry.

Between February and August 2000, the number of Napster users rose from 1.1 million to 6.7 million, making it the fastest-growing software application ever recorded.[54] The program was used by an estimated 6 percent of U.S. home-PC users with modems; Napster claimed that 28 million people had downloaded its program.[55] On May 19, 2000, the venture-capital firm Hummer Winblad (which also funded Liquid Audio) agreed to pump between $15 million and $20 million into Napster. Although Napster had no revenue streams in the form of ads, subscriptions, or database trafficking (the holy trinity of e-commerce), the system's millions of registered users constituted a substantial audience and a potential subscriber base. The larger the connected base, the greater the value of the network, as users provided a vast reservoir of content that increased with the network's popularity.

Napster participants did not sell their free copies and did not profit by their copying. Although Napster wasn't making money off its users, the Big Five argued that Napster's users weren't making any money for them. On June 12, 2000, the RIAA filed a motion for an injunction against Napster based on a study of 2,555 college students by Field Research Cooperation, which showed a direct correlation between Napster use and decreased CD sales. The study, based on SoundScan sales data, reported that although overall CD sales steadily increased from January 1997 to March 2000, sales near campuses "where anecdotal evidence suggested a high degree of Napster usage" dropped 7 percent.[56] Relying as it did on anecdotal evidence rather than a scientific sampling method, the study was fundamentally unreliable. Competing anecdotal evidence soon indicated that users used Napster not to "steal" music but, as they used radio, to "sample" it before purchasing; they were drawn to the huge array of music it offered, much of which was inaccessible by any other means.[57] Moreover, the single tracks that Napster offered—traditionally a small part of retail-store business—did not substitute for CDs. CD retailers' sales were obviously much affected by new distribution trends, such as the growth of big chain stores and online retailers such as CDNow, and competing products like video games. Indeed, CD shipments in the United States rose 10 percent in 1999, even as levels of MP3 file sharing rose.[58]

Artists were divided over the impact of Napster on music sales. In April 2000, the rock group Metallica charged that Napster, along with Yale University, Indiana University, and the University of Southern California (USC), violated copyright laws by enabling students to swap digital music files, and sought $10 million in damages. Yet the band had benefited earlier from providing free content to fans; the *New York Times* noted that

> any rift between Metallica and its fans over Napster would be particularly striking since the band's initial popularity in the early 1980's grew in part from a cassette demo tape, "No Life Til Leather," that the band sent out free and encouraged fans to bootleg. It is said to be among the most-traded cassette demo tapes in rock.[59]

On May 3, 2000, Metallica appeared on Napster's doorstep with a list of 335,435 users who had posted sound files of the band's works. The list was compiled by an English-based company, MP3 Police, who had contacted Metallica with an offer to hunt down Napster users suspected of infringement. Napster promised to bar anyone who had infringed copyright, yet users could easily reregister under different names. Rapper Dr. Dre also filed a lawsuit against Napster, but other artists rallied to its defense, including Limp Bizkit (who received tour funding from Napster), The Offspring, and Public Enemy, whose leader, Chuck D., remains one of the most virulent critics of the major labels.

Legal Arguments and Injunctions

In its defense, Napster claimed its system allowed users to "space shift" or transfer their MP3s between hard drives and players, just as taping had enabled consumers to transfer their music from LPs to cassettes; many users, Napster argued, used its system simply to access music that they already owned on CD from remote locales like a dorm room. Napster invoked the 1992 Audio Home Recording Act, which granted consumers the right to transfer digital music for personal, noncommercial use, and *RIAA v. Diamond Multimedia Systems* (1999), in which the court ruled that the AHRA allowed all noncommercial copying. The RIAA countered that the AHRA applied only to manufactured products such as portable music players and not to software applications; computer hard drives, it claimed, were not digital-recording devices under the act.[60] A friend-of-the-court brief filed by the U.S. federal government supported the RIAA's position.

Napster also cited the Supreme Court's five-to-four ruling in 1984's *Sony Corp. v. Universal Studios*. In the Betamax case, as it is popularly known, the Court ruled that the provider is not liable if the technology is "capable of substantial non-infringing uses." Napster's lawyers argued that "Napster cannot be guilty of vicarious or contributory infringement, because the service unquestionably involves substantial noninfringing uses" such as space shifting and distribution of noncopyrighted music.[61] The RIAA countered that *Sony* referred to private copying at home rather than to public exchanges of anonymous users via the Internet, which was unprecedented in scale.[62] The RIAA also noted that Napster had an ongoing relationship with its customers that Sony lacked; therefore, Napster could work to prevent illegal copying. The RIAA also claimed that Napster users had a reduced claim of privacy, since they were making their hard drives available to a public forum. The Betamax analogy proved problematic also because of its basis in patent law: "While if the Court had ruled against Sony, an entire technology would have been taken off the market, Napster has no patents and relies on generic and widely used database and file-transfer software in its operations, programs that will continue to exist even if Napster shuts down."[63]

In the end, Napster cited the First Amendment, claiming that "just like a magazine or a newspaper that provides information—that provides an index or a directory—Napster has the right to provide an index, a directory. It's a dissemination-of-information right. And the courts have held that directory publishers enjoy free speech rights."[64] An injunction against Napster would amount to prior restraint on the speech of Napster's users/publishers. Its attorneys also claimed that sharing files with strangers constituted fair use, since there was no commercial gain for providers. However, "Most copyright experts say the fair-use doctrine applies to personal use within a household, not to sharing music files with hundreds or even thousands of strangers via the Internet."[65] Videocassette recorders are not a digital, network-based technology. In *Sony*, the Court didn't define "limited number," but this proviso had since been interpreted as applying only to personal use.[66]

In late July 2000, federal judge Marilyn Patel ordered an injunction against Napster, rejecting the fair-use and "substantial noninfringing uses" arguments and finding instead that Napster was used primarily to download copyrighted music.[67] Judge Patel rejected Napster's claim that under the 1998 DMCA, it was a "mere conduit" for digital information, like an Internet-service provider, and couldn't be held responsible for user behavior. Patel suggested that Napster had failed to post a written policy describing how it would deal with copyright infringement, which would have protected it under DMCA. The injunction would not have shut Napster down entirely; it only covered music copyrighted by the plaintiffs in the case. Napster could still operate its chat rooms and offer downloads of artists who had approved their distribution in its "New Artists Program."

Napster then hired David Boies, the Justice Department's special counsel in the Microsoft antitrust case. Judge Patel's ruling was stayed by an appeals court, which found that Napster wasn't given time to offer a detailed explanation of how its technology worked or to provide additional evidence regarding its noninfringing uses. Most significantly, studies on Napster's effects on sales (which were not considered by Judge Patel) dispelled the RIAA's claims on usage and piracy. The role of college students was overestimated; a Pew study conducted in April 2000 found that, "of the 13 million American 'freeloaders,' only 37 percent are college students. In fact, fewer than half (48 percent) are between the ages of 18 and 29. Forty-two percent are between 30–49, while the remaining 10 percent are over 50."[68] A Student Monitor study showed that Napster was only the fourth most common music site visited by students, and "only 17 percent reported using the service to download free music."[69]

The BMG Dalliance

In the summer of 2000, Napster and the Big Five entered into talks to settle the lawsuits. Hank Barry, Napster CEO, stated, "We have no desire to put the labels out of business."[70] On October 31, shortly after the Napster decision, BMG announced that it would loan Napster $50 million to develop a secure file-sharing system that would "preserve the Napster experience" while compensating copyright holders. When such a model was developed and operating, BMG would no longer pursue its lawsuit against Napster.[71] In exchange, BMG retained the right to take a 58 percent interest in Napster when the new service was developed.[72] However, as BMG had joined the other Big Five companies in suing Napster, it also was suing itself. What accounts for such logic? While publicly claiming that no legal means for online music sharing existed, the Big Five privately hedged their bets through mergers and acquisitions that would allow file sharing under their exclusive control.

In particular, BMG was attracted by Napster's brand recognition, tangible assets, and delivery software (including the protocol and interface). If all went as planned, BMG would have content, distribution (Napster and CDNow), and marketing (GetMusic.com). A direct link from the Napster interface to the CDNow website was seen as a means to increase CD sales; as one observer noted, "Bertelsmann-owned CDNow has also become the de facto retailer for the Napster service, with a purchase button included in the current application."[73] BMG also could take a chance on Napster because it could be more flexible with its capital than its competitors could. It was the only privately held Big Five company, and it was not tied to a film studio with potentially cash-draining flops. BMG saw Napster as a win-win situation: if the court ruled in favor of Napster, BMG would not press its lawsuit, while if the court ruled against Napster, BMG was not liable for damages.

"Napster II" was scheduled for introduction in June 2001, based on a three-tiered system. The first tier would use viral marketing, allowing nonsubscribers to trade files with a lifespan limited to two months for free. The files would be encrypted and encoded at a low bit rate, and users couldn't transfer downloads to other players.[74] A second tier would charge a $4.95 monthly "membership" fee for access to limited downloads, and the third tier would charge additional per-song fees for access to wider content. Napster II would also incorporate digital-rights management technology. Files would be encrypted when uploaded to the central server. They then would be forwarded to the downloader, who would have a "key" that unlocked the encryption and allowed the authenticated files to be played. According to Mann, "In Napster II, songs would fly around the service almost as freely as before, but any particular copy of a song would be usable only by a single Napster account."[75]

Napster II faced many problems, however. By charging subscription fees for a formerly free service, Napster II would enter a vicious cycle and drive off an estimated 40 percent of users.[76] Fewer users would mean diminished offerings, and that in turn would reduce incentives to subscribe. While BMG obviously had a stake in encouraging other companies and publishers to join up, these companies and publishers were not interested in joining a system in which BMG had controlling interest (the largest of the Big Five, Universal, had been a particularly hard-line opponent of Napster). The software for Napster II encryption also was problematic. Encryption would take place on users' PCs rather than on the central server. This arrangement would be convenient for Napster, yet it promised to slow transmission and complicate use.

BMG's plans were dashed when a three-judge panel unanimously upheld Judge Patel's injunction in February 2001, ruling that Napster "knowingly encourages and assists in the infringement of copyrights."[77] The panel found that Napster had a direct financial interest in its users' infringing activity and that it neglected its ability and duty to police its system for infringing activity. The panel rejected the fair-use and AHRA defenses and disallowed presentation of evidence purporting to show that Napster use was linked to increased CD sales. The judges did find that the scope of the original injunction was overbroad, noting that the service was capable of "commercially significant noninfringing uses." Napster did not have to shut down; it merely had to block access to files of copyrighted material. Although ostensibly a victory for the Big Five, the decision placed them in the position of having to constantly monitor and notify Napster of infringement in what Lawrence Lessig termed a "war of attrition."[78] On March 5, Judge Patel ordered Napster to remove copyrighted files within three days of notification by record companies.[79]

Anxious to settle with the recording industry, Napster offered $1 billion in payments over five years to the Big Five, who would share $150 million annually from subscription fees. Another $50 million would be split annually between independent labels. Ever optimistic, Napster predicted that its revamped service would have 2.5 million subscribers by the end of the year, and 8.5 million by 2005, with annual revenues growing from $149 million to $505 million during that period.[80] The Big Five rejected the offer even before it was made public. One analyst said that $200 million per year is "roughly the same as what the CD singles business is worth. . . . This settlement doesn't even begin to address the potential drop-off in the album market."[81]

Napster's Death and Reincarnation

Napster use dropped 36 percent between March and April 2001, and during this period, the average number of songs offered per user dropped from 220 to 37.[82] By late June, only 320,000 users were logged on at any given time, and they were offering only 1.5 songs apiece, down from February's peak of 1.57 million simultaneous users offering 220 songs apiece.[83] On July 3, 2001, Napster blocked file transfers altogether and shut down its operations. Ten days later, the company settled lawsuits with Metallica and Dr. Dre for an undisclosed amount. In September, Napster agreed to pay $26 million to settle lawsuits with songwriters and music publishers, and an additional $20 million as an advance on future licensing fees—fees paid by BMG, which by that time had sunk $100 million into the company.[84]

In January 2002, four of the Big Five suspended their lawsuits against Napster (EMI declined on grounds that it was pursuing settlement talks). Their containment strategy had boomeranged; by refusing to license recordings to Napster, they appeared guilty of collusion in furthering plans for their own subscription services.[85] Judge Patel wrote, "These ventures look bad, smell bad, and sound bad. If Napster is correct, these plaintiffs are attempting the near monopolization of the digital distribution market."[86] Inquiry into their practices could have undermined the whole "works-for-hire" basis of the recording industry. As one observer noted, "Musicians are in a gray area. Since they sign contracts and don't draw a base salary, it's unclear if they are freelancers who retain ownership or employees who forfeit their rights."[87] By questioning basic issues of control over copyrights, Napster's attorneys could have done to the Big Five what the Big Five did to Napster.

On May 17, BMG agreed to pay Napster's creditors $8 million to acquire the company's assets. Napster filed for bankruptcy on June 3, listing $7.9 million in assets and $101 million in liabilities (including $91 million in loans from BMG).[88] However, a judge blocked the sale on grounds of conflict of interest; Napster CEO Konrad Hilbers, formerly of BMG, had "one foot in the Napster camp and one foot in the Bertelsmann camp."[89] BMG washed its hands of Napster after the departure of CEO Thomas Middelhof, and what was left of the company (essentially its "intellectual-property profile," including software code base, domain name, and logo) was acquired by Roxio, a multimedia software vendor, for $5 million and 100,000 stock warrants in November.[90] Roxio was a stakeholder with Universal and Sony Music in their Pressplay music service, which used the company's CD-burning technology.[91]

Only three years after the demise of Napster, a software firm relaunched Napster in a new commercial format, and the program that once filled administrators of college and university networks with legal angst is now, at some campuses, given compulsory subsidies from students, who receive Napster as a paid service.[92] The culture industry redeemed Napster, transforming it from pariah technology to college requirement. In the process, Napster II changed its business model and its architecture, offering subscription services operating on client-server software and a "to-go" subscription service for customers with mobile MP3 players, who lease time to access the online catalog. Nonetheless, the spirit of the original Napster lives on in the free-software community. OpenNap, an open-source server software package, creates a peer-to-peer file-sharing community using the Napster protocols without a centralized server.[93] JNerve and SlavaNap are other Napster-ish efforts, and open-source applications derived from Napster include the iNapster Web interface, the Hackster Visual Basic Napster client, and the WebNap PHP Napster client. Rather than focusing on a single application, peer-to-peer systems have now diffused thoroughly throughout the Internet.

THE PERSISTENCE OF PEER-TO-PEER

The Napster decision eliminated the legal basis for a not-for-profit alternative to the Celestial Jukebox, but peer-to-peer technology continues to disrupt the industry. By July 15, 2001, when Napster shut down, various P2P services logged nearly 5,850,000 downloads.[94] Although its legal machinations had snuffed Napster, the industry was unable to snuff the file-sharing practice to which Napster had introduced millions of users. New services structured themselves to circumvent Napster's fatal flaw: a centralized directory that kept track of files. These services eliminated the client-server relationship, allowing individual users to exchange video, music, and text files without passing these through a central server.

The recording industry quickly placed these new and "unauthorized" services in its sights. Aimster, a system that allowed AOL Instant Messenger (AIM) users to search their "buddies'" hard drives for text, audio, and video files, was launched in the summer of 2000. Aimster integrated Napster protocols to enable AIM users to set up file-swapping networks in which indexes of files existed only on the hard drives of AIM buddy-group members, thus essentially creating a system of closed communities, or "small-worlds" networks.[95] Emboldened by their success against Napster, the Big Five filed suit against Aimster in May 2001. Aimster claimed that it should be exempt because it allowed users to trade files through private message exchanges; its founder, Johnny Deep, said Aimster was "'not a music sharing service,' but a private network used to exchange all kinds of information. He said it was not the 'right or responsibility' of Aimster to track what sorts of files its users exchange."[96] A month later, Columbia Pictures, Disney, MGM, Paramount, Sony, Twentieth

Century Fox, and Universal joined the suit. A court ruling halted the operation, and Aimster (now Madster) filed for bankruptcy in March 2002. In 2003, a U.S. appeals court upheld an injunction preventing Aimster's distribution.[97] In January 2004, Aimster/Madster lost a bid to have the U.S. Supreme Court appeal the ruling that shut it down.

Freenet and Gnutella

In the wake of Napster, much industry and media attention was focused on Freenet, a peer-to-peer system developed by Ian Clarke, a twenty-three-year-old Irish programmer who posted a test program on the Internet in March 2000. Once Freenet is installed on a computer, that computer becomes a Freenet peer, routing received messages to their destinations and sending and receiving data about its traffic. Each file is encrypted and distributed among the people using the program, who don't know the contents of files: "Freenet incorporates a digital 'immune system' that responds to any effort to determine the location of a piece of information by spreading the information elsewhere in the network."[98] As it distributes encrypted data packages, no one but the requesting user can "look into" the contents of the data streams entering or leaving a user's computer. Location and source are impossible to determine; file transfers are untraceable. One observer noted that "Clarke's motives are political—his dream is to liberate intellectual property. 'My opinion is that people who rely on copyright probably need to change their business model,' he says."[99] Despite these noble goals, a November 2000 survey of Freenet activity found that 15.6 percent of all Freenet activity was devoted to pornography, and 53.8 percent of Freenet text, audio, and video files was devoted to the Devil's triumvirate of sex, drugs, and rock and roll.[100] Despite the cornucopia of wonders Freenet offered to the troops in Teenland, its adoption rate was limited. Freenet was difficult to install and use, slow to locate material, and unable to perform keyword searching.

Gnutella, which serves as the basis for file-sharing programs like LimeWire and BearShare, originated with a code written by Justin Frankel, a programmer at Nullsoft, which was owned by AOL Time Warner. Frankel's corporate patrons at AOL, outraged by the fact that he had developed the program on the company dime while they were busy battling the Napster scourge, suppressed the project. Its development was then picked up by a "global network of programmers, who collaborate in a fashion similar to the development of the free Linux operating system."[101] Unlike Napster, Gnutella is not owned by anyone; it is an open protocol for which anyone can write a client application. The program was described as follows:

> Users . . . in essence form a search engine of their own that expands its search exponentially. When a Gnutella user has a query, the software sends it to ten computers on the network. If the first ten computers don't have a file, each computer sends it to ten other computers and so on until [theoretically] an estimated million computers would be looking for it in just five to ten seconds. The program could theoretically check every site on the Web.[102]

Yet Gnutella faced problems that plague all decentralized systems. The first has been termed the "Gnutella paradox, [in which the] attainment of widespread popularity may in fact signal a file trading software program's imminent demise."[103] A single query could produce a cascade of network traffic that overloaded many peers and created a "mesh of dead ends."[104] The program was revised to create hubs, or "reflector nodes," with a broadband Internet connection that could form a flexible backbone for efficiently handling network traffic. Another problem arose from the "tragedy of the Digital Commons," a consequence of freeloading and lack of participation among users. A study of Gnutella use found that "about 25 percent of the Gnutella users serve about 98 percent of the files," 1 percent of users provide 40 percent of the files, and 2 percent of users over half.[105]

In general, users for peer-to-peer networks have learned to fear viruses, worms, and harmful programs. Distrust and disruption will break down a network once it gets large enough: "Balancing the privacy of the individual with the need to authenticate users and kick malicious disrupters off the system is a difficult feat."[106] Users must take care opening their computer's ports to P2P traffic, as cookies residing on hard drives may be appropriated by other users for illicit commerce, and other areas not cordoned off by users may be freely examined. There is a fundamental contradiction between security issues, which work against expanding community access, and peer-to-peer networking, which requires such expansion.

FastTrack

Gnutella's principal rival, FastTrack, was founded in the Netherlands in April 2000 by a group of programmers from Sweden, Denmark, and the Netherlands, headed by Nikola Zennstrom. FastTrack became the most common model for peer-to-peer systems, used by KaZaA, Audiogalaxy, MusicCity Morpheus, WinMX, Grokster, and others. FastTrack earned revenues from licensing software to these systems, and more than $1.4 million from "adware" companies that attach plug-ins to KaZaA (also founded by Zennstrom). It addressed a principal problem of peer-to-peer systems, scalability, by limiting redundant search requests. Its system, a compromise between a central server and a truly decentralized peer-to-peer system, employed supernodes—users with the most powerful computers and greatest bandwidth at any given time—which the software was able to seek out and turn into temporary search hubs. MusicCity Morpheus limited the number of files a user could share and created unique identifiers, or "hashes," for files. Morpheus works as follows:

> A user's Morpheus client transmits its Internet address and names of shared files to a single supernode in an encrypted form; in turn, about every 10 minutes the supernode tells the client how many users and files are available on the whole network, Webnoize found. When a client requests a search and then asks for a file, the supernode searches its own directory and passes on the request to other supernodes. . . . [Once] a supernode finds the file, it sends the client encrypted addresses and matching file names, allowing the client to contact another client directly and begin downloading. Webnoize found that the peer-to-peer download process itself isn't encrypted, so the file transfer can be observed by anyone along the transmission path with the appropriate equipment.[107]

The early leader in post-Napster services, KaZaA, was purchased in January 2002 by an obscure Australian multimedia company, Sharman Networks Ltd., for approximately $500,000.[108] Sharman illustrates the global and slippery nature of ownership in the Digital Age. According to the *New York Times*, the company was "incorporated in the South Pacific island of Vanuatu and managed from Australia. Its computer servers are in Denmark and the source code for its software was last seen in Estonia."[109] FastTrack and Zennstrom, KaZaA's founder, were charged with copyright infringement shortly after KaZaA's release, but in March 2002, a Dutch appeals court ruled that they weren't liable for copyright infringement by users of FastTrack's applications.

KaZaA owed its growth to the demise of Napster in mid-2001, and to Sharman's disconnection of its chief competitor—Morpheus's creator, StreamCast Networks—from the FastTrack network. Despite participating in production and distribution of P2P file-sharing software, Sharman was not moved by any altruistic desire to "share the music." Shortly after its acquisition, Sharman loaded KaZaA with adware and altered the software to include a Trojan-horse program that would steer users to a second network that charged fees for access to files (Brilliant Digital Entertainment's Altnet).[110] As one report explained,

> Why does Sharman do it? Because the more users, the better. A bigger consumer base allows Sharman to sell more ads and to devise new revenue opportunities—like the complex scheme Kazaa (and other file-sharing services) tried to deploy last year. It is hijacking commissions from e-commerce sites like Amazon that were earmarked for referring organizations (everything from blogs to nonprofit sites), and diverting them to a third party, which in turn paid Sharman.[111]

On October 2, 2001, FastTrack services were sued by the Big Five and several major Hollywood studios. *Metro-Goldwyn-Mayer Studios v. Grokster et al.* amounted to a "10-foot stack of papers" that accused FastTrack-based services KaZaA, StreamCast, Morpheus, and Grokster (another offshore entity headquartered in Nevis, West Indies) of operating a "21st-century piratical bazaar."[112] Specifically, the recording and film industry claimed that these services promoted "secondary" and "vicarious" copyright infringement. Peer-to-peer services scored a major victory on April 25, 2003, when Federal Judge Stephen Wilson dismissed the suit against StreamCast and Grokster, stating, "Defendants distribute and support software, the users of which can and do choose to employ it for both lawful and unlawful ends. Grokster and StreamCast are not significantly different from companies that sell home video recorders or copy machines, both of which can be and are used to infringe copyrights."[113]

In stark contrast to the judge in the Aimster suit, Wilson accepted the "Betamax"-styled argument that the owners of software cannot control user behavior. In an appeal of the Grokster decision in the Ninth U.S. Circuit Court of Appeals in February 2004, the RIAA and MPAA argued that the Betamax decision was irrelevant to Grokster, because peer-to-peer systems could readily identify files and also earn revenues from advertising to network users. The Electronic Frontiers Foundation (EFF) countered that much of the content was legal and that, moreover, the legality of content was irrelevant because the networks could not monitor or prevent the activities of their users, just as the makers of CD burners could not control whether they were used to burn legal or illegal CDs.[114] Copyright and culture continue to coexist uneasily in cyberspace.

The Recording Industry's Response

The decision by the federal court in favor of Grokster and StreamCast, and the continuing popularity of peer-to-peer services, forced the RIAA to reconsider its tactics. It mobilized a public-relations campaign of fear and doubt. Record executives accused peer-to-peer systems of spreading child pornography. Andrew Lack, the CEO of Sony Entertainment, huffed, "As a guy in the record industry and as a parent, I am shocked that these services are being used to lure children to stuff that is really ugly." The *New York Times*, however, found that "file sharing makes up a small and shrinking portion of all reported child pornography," adding, "Others may ask whether raising this issue is more than a little cynical from an industry that heavily promotes music with sexual and violent themes."[115]

The industry also promoted "spoofing": posting to peer-to-peer services decoy files that contain noise, static, or loops of a song segment in order to frustrate users trying to download these songs. The industry also attacked peer-to-peer networks through "redirection," in which a posted file automatically redirected its recipient to a commercial website or a warning message, allowing the sites to obtain these users' Internet addresses in the process. And it considered, but never tried, yet another method of attack, "interdiction," which would overload a computer sharing files with a flood of download requests.[116] By July of 2003, navigating peer-to-peer systems had increasingly become an exercise in frustration:

> They are strewn with poor-quality music files and intrusive spyware that monitors users' movements and bombards them with pop-up ads, along with decoy files that artists and record companies put out to trick downloaders. For many, the idea of spending time negotiating through peer-to-peer networks is pure hell.[117]

Yet the systems remained popular, and some in the government called for more draconian measures. Orrin Hatch, head of the Senate Judiciary Committee, announced in June 2003 that he favored developing ways to death zap the computers of peer-to-peer users: "If that's the only way [to stop illegal online behavior], then I'm all for destroying their machines."[118] In July 2002, the RIAA announced a change of strategy: rather than suing software companies, it would sue the users of peer-to-peer services, specifically the uploaders who provided the bulk of files to the services. While peer-to-peer networks veil the identities of users behind the nodes, they reveal the username and IP address associated with each node:

> In order to pursue apparent infringers the RIAA needs to be able to identify the individuals who are sharing and trading files using P2P programs. The RIAA can readily obtain the screen name of an individual user, and using the Internet Protocol (IP) address associated with that screen name, can trace the user to his ISP [Internet service provider]. Only the ISP, however, can link the IP address used to access a P2P program with the name and address of a person—the ISP's customer—who can then be contacted or, if need be, sued by the RIAA.[119]

When the major record companies began to hint at suing participants in peer-to-peer networks, Warner Music Group chairman Roger Ames refused to join in until the other companies agreed to sell their catalogs on MusicNet and Pressplay.[120] The Big Five tested the waters in May 2003, when four university students who were sued for operating file-sharing programs on their school networks settled with the RIAA for between $12,000 and $17,500 each.[121] In July, the RIAA obtained nearly one thousand subpoenas against suspected uploaders, demanding their names from service providers and universities. Under copyright law, the RIAA could be awarded damages of $750 to $150,000 per copyrighted song.[122] In early September, the RIAA offered an amnesty program in which suspected uploaders would be spared prosecution if they signed an affidavit, accompanied by a photo ID, promising to delete illegal copies of songs from their computers and desist from future sharing, after which their names would then be entered into a database. The logical inconsistencies of the "amnesty plan" were obvious. As one observer noted, "That's like saying, 'Come tell us if you have any intention of becoming a revolutionary.'"[123]

Unsurprisingly, response to the amnesty program was underwhelming, and on September 8, 2003, the RIAA filed suit against 261 people they accused of illegally making copyright material freely available by uploading it to file-sharing programs such as KaZaA. One suit that attracted a substantial amount of media attention involved a twelve-year-old honor student, Brianna Lahara, who lived in a Manhattan housing project. The family had paid $29.99 for the KaZaA service three months earlier. When informed that she was being sued, Brianna told the *New York Post*, "I thought it was OK to download music because my mom paid a service fee for it."[124] The RIAA settled the suit with Brianna for $2,000. Her mother, Sylvia Torres, told the *Wall Street Journal*, "You can be sure Brianna won't be doing it any more."[125]

Another target was Sarah Ward, a sixty-six-year-old retired schoolteacher who allegedly made thousands of songs available through KaZaA. The RIAA said it had gathered evidence showing Ward had used KaZaA to share more than two thousand songs, including hip-hop hits like Trick Daddy's "I'm a Thug."[126] Ms. Ward professed to being deeply confused by the RIAA suit, stating that she did not trade music or have any younger people living with her. "I'm a very much dyslexic person who has not actually engaged in using the computer as a tool yet," she told the *New York Times*.[127]

Clearly, the RIAA was not particularly careful in choosing whom to sue, except insofar as it sought to protect its own members: it sued KaZaA users overwhelmingly; subscribers to AOL (owned by Time Warner) were left alone.[128] Although Sarah Ward escaped penalties, a U.S. House representative from Texas, John Carter, promoted a Texas-style tough-justice bill requiring prison time for all instances of illegal file sharing.[129] Other get-tough tactics of the RIAA were presented in so-called "public-education" campaigns that were widely covered in uncritical news reports.

In December 2003, a Washington, D.C., appeals court ruled that the RIAA could not force ISPs to reveal the identities of alleged uploaders and therefore could not file a single lawsuit against large numbers of suspected violators. Chief Judge Douglas Ginsburg, writing for a unanimous three-judge panel, ruled that Verizon, the ISP in the case, was merely serving as a conduit for copyrighted material rather than hosting it on its computers in violation of the DMCA.[130] The RIAA then filed "John Doe" suits against individuals, naming only the Internet Protocol address that they believed was attached to a computer engaged in illegal activity. If the judge considered the suit worthy of consideration, he or she would grant a subpoena, and the ISP would then be forced to reveal the user's identity.[131] By September 2004, the RIAA had filed suit against 4,679 people for distributing songs on peer-to-peer networks. Of these suits, 1,024 were settled, paying the RIAA an average of $5,000.[132]

The effect of the lawsuits is uncertain. In March 2004, the Pew Internet and American Life Project found that 14 percent of Americans with Internet access claimed to have stopped downloading music files.[133] According to online tracking firm BigChampagne, however, 8.3 million people were using peer-to-peer services at any one time in June 2004, an increase of 19 percent from the 6.8 million using the services during the same period a year earlier.

KaZaA, however, was in decline. It continued to shed users throughout 2004, dropping from an October 2003 peak of 5.6 million users to 3.8 million eight months later.[134] In addition to bearing the brunt of the suits almost exclusively, KaZaA users were plagued by viruses, pop-up ads, and spyware, and they switched to newer services such as eDonkey and BitTorrent. The latter, created by Bram Cohen and released in October 2002, represented a third generation of peer-to-peer services in the wake of Napster and noncentralized systems like Gnutella. The BitTorrent system cuts up files into sections, which, as they are downloaded by one user, become available through that user to other users, "[so] almost all of the people who are sharing a given file are simultaneously uploading and downloading pieces of the same file" unless downloading is complete.[135] This system enables rapid sharing of large files to substantial numbers of people while placing minimal bandwidth requirements on the original provider. The ingenuity of file-sharing "hacktivism" seems likely to stay a step ahead of the RIAA reformers, and it helps music fans in the information society keep the "record" button on their computers that the music industry keeps trying to take away.

File Sharing's Impact on Recording Sales

A report prepared for the recording industry predicted that by 2002, an estimated 16 percent of all U.S. music sales, amounting to $985 million, would be lost to file sharing.[136]However, some executives were unconvinced. Jay Samit, senior vice president of EMI, said, "We've far more to fear from a surplus of CD manufacturing here in

Asia, where in some markets 90 percent of CDs are bootlegged, than from the Internet."[137] In many respects, the industry's failure to make peace with Scour and MP3.com, the Napster crackdown, and the post-Napster vilification of peer-to-peer systems were compounding miscalculations that have cost the industry dearly, both in missed sales opportunities and in lost goodwill with music fans. Early analyst research suggested that Napster use spurred CD sales, rather than suppressing them. While the music industry claimed to lose money to downloading, a study of the Napster period found that the majors released 25 percent fewer titles in 2001 than in 1999. At the same time, the average price of a CD rose 7.2 percent—during a period of negligible overall economic inflation.[138] In fact, recorded-music sales in the United States reached an all-time high in 2000. Overall sales (785.1 million units) were up 4 percent from 1999 (754.8 million units).[139] In one week, ending December 24, 2000, 45.4 million CDs were sold; sales were up 13.6 percent from the same period in 1999.[140]

A host of factors, when taken together, explains the subsequent drop in CD sales as plausibly as does file-sharing piracy.[141] At the turn of the century, disposable income declined in a recessionary U.S. economy, and the music industry faced growing competition from other media, such as video games and DVDs. The Big Four also ignored substantial markets by favoring teenagers.[142] Consolidation within the radio industry (and MTV's growing neglect of music videos) resulted in shorter playlists, and shrinking retail space created greater barriers to entry for new releases and less exposure for money-earning catalog recordings. The quality of releases also played a role: analysts noted that a lack of "superstar" releases diminished overall sales. In 2001, for the first time in thirty-five years, no release sold five million copies.[143] Most importantly, the "units-shipped" figures cited by the RIAA to bolster its antipiracy arguments cannot in fact serve as a reliable index of declining sales. The RIAA reported a drop of 74.1 percent in CD-single units shipped in 2001–2002 and implied that file sharing was to blame, yet fewer singles were sold because fewer were pressed: the Big Four were abandoning them on the grounds that they cut into the sales of more profitable full-length CDs. Only seventeen of the top fifty songs on the March 3, 2000, *Billboard* "Hot 100" were available as singles.[144]

A Yankelovich poll released in June 2000 reported that 66 percent of all consumers who had downloaded music "said that listening to a song online has at least once prompted them to later buy a CD or cassette featuring the same song."[145] In June 2000 (a month after USC banned Napster), the Annenberg School at USC released a survey that found that 63 percent of students who downloaded MP3s still bought the same number of CDs, and 10 percent bought more CDs; 39 percent of students who downloaded MP3s said that after listening to them, they bought the corresponding CDs for their superior sound quality.[146] A June 2001 survey of three thousand adult peer-to-peer users by Jupiter Media Metrix found that they were more likely to buy new music than were average music fans. The author of the report, Aram Sinnreich, stated, "File-sharing is a net positive technology" in promoting sales. The study found that 34 percent of experienced file sharers had decreased their spending on music, and that 52 percent of experienced file sharers had increased their music purchases. "By comparison, among average Internet users who describe themselves as music fans, and who may or may not use file-sharing networks, only 19 percent said their spending increased. Roughly 70 percent of this group found that their spending had stayed constant."[147]

A study by Oberholzer-Gee and Strumpf used direct data, rather than surveys, of download activity in fall 2002 and compared these with music purchases in the same time frame. "While downloads occur on a vast scale, most users are likely individuals who would not have bought the album even in the absence of file sharing."[148] The study was conducted by monitoring 1.75 million downloads over a seventeen-week period; examining server logs from an open-source Napster server, OpenNap; and comparing sales of nearly seven hundred albums from Nielsen SoundScan. The study indicated that most downloading is done by teens and postteens, who are "money-poor but time-rich" and otherwise would not have purchased the recordings they downloaded. In contrast, file sharing actually increased sales among older users, who used peer-to-peer systems to "sample" recordings before deciding to purchase.[149] In an interview with *Working Knowledge*, Dr. Oberholzer-Gee stated, "Our research shows that people do not download entire CDs. They download a few songs, typically the hits that one would also hear on a Top 40 station. This suggests that P2P is much like radio, a great tool to promote new music."[150] File sharing was a community-building ritual that the majors could have turned to their advantage—a missed opportunity for a more humane Celestial Jukebox.

The Darknet and the Future of Peer-to-Peer

The RIAA's claims linking lost revenues to illegal file sharing may have been debunked, but the recording industry has argued in the Verizon case and in press releases that peer-to-peer systems pose an immediate and absolute danger to its survival. "Worldwide, the recording industry has shrunk from a $40 billion industry in 2000 down to a $32 billion industry in 2002. . . . In 2000, the ten top-selling albums in the United States sold a total of 60 million units. In 2001, that number dropped to 40 million. . . . The root cause for this dramatic decline in record sales is the astronomical rate of music piracy on the Internet."[151] The Big Four continue to lobby for legislation such as the "Inducing Infringements of Copyright Act" (or "Induce Act"), which would outlaw the creation and use of P2P technology (or any technology that makes copies).[152] Although its U.S. Senate sponsor, Orrin Hatch, subsequently withdrew the first introduction of the Induce Act bill, it undoubtedly will resurface in the future. The Electronic Frontiers Foundation critiqued the intent and the consequences of the Induce Act:

> Apple's iPod music player seemed particularly vulnerable to attack. Any major record label could bring a strong lawsuit against Apple for "intentionally inducing" infringement under this new law with the iPod, both because it's plausible to argue that having an iPod enhances the lure of using P2P to download music (gotta fill all that space!) and because all the major record labels still believe that private sharing of songs from your CDs with friends is copyright infringement.[153]

The file-sharing networks that blossomed with nurturing from venture capitalists until the Napster shutdown created an alternative service to the Celestial Jukebox before the "official" Jukebox was ever built.[154] In fact, the industry's MSPs were created to fulfill its need for a formal, legal rebuke of the communal P2P technologies most used for MP3 file sharing, including Napster and Audiogalaxy, and to turn away once and for all from dalliances with an alternative jukebox.

Disruption is in the eye of the beholder, and, like Universal, MP3.com, BMG, and Napster, some beholders have partially embraced disruption. In January 2001, Sony introduced a line of CD players that also played MP3 files burned onto CDs. As one reporter noted,

> It's an about-face as abrupt as record label BMG's embrace of Napster. When asked to describe Sony Music's reaction, one person at Sony Electronics said sheepishly, "They were pissed." As Sony Electronics sees it, they had no choice, given the tortoise-like pace of industry-sanctioned digital music compared to Napster's runaway success.[155]

Another example of corporate dalliances with the Darknet that created mixed signals involves BMG, which partnered with Grokster to distribute licensed music, while simultaneously suing Grokster as a party to *MGM v. Grokster*. In short, the Big Four responds to peer-to-peer networking by vilifying and prosecuting it, but also by investing in it through joint ventures. They are hedging their bets.

The recording and film industries have attacked peer-to-peer networks with blunt legal instruments, seeking in particular to defeat the argument that such networks have substantial noninfringing uses (the legal equivalent of software free speech). This argument protects e-mail from legal challenge. As Biddle and his colleagues note, "[E-mail] has substantial non-infringing use, and so e-mail itself is not under legal threat even though it may be used to transfer copyrighted material unlawfully." Peer-to-peer file-sharing networks incontestably have substantial noninfringing uses, and other applications include free telephony (Skype VoIP), workgroup collaboration (Groove), protection from viruses (Rumor), storage (Centrata),[156] and scientific research (SETI@Home).

"Trusted computing" is industrial jargon for the lockdown of business and cultural flows on the Internet through digital-rights management and its supporting technologies. As Rosenblatt and Dykstra explain, "Adding *persistent protection* to content is the most effective way to control and track access."[157] Yet these "persistent protection[s]" have very real costs to individuals and society:

> They provide security to users while giving third parties the power to enforce policies on users' computers against the users' wishes—they let others pressure you to hand some control over your PC to someone else. This is a "feature" ready-made for abuse by software authors who want to anticompetitively choke off rival software.[158]

These possibilities for removing users' control are implicit in a document titled "The Darknet and the Future of Content Distribution,"[159] which played an instrumental role in the design of the Celestial Jukebox. The Darknet paper, which became influential after its authors (employees of Microsoft Corporation) published it on a Stanford website, defines the Darknet as "a collection of networks and technologies used to share digital content" that requires "an application and protocol layer riding on existing networks" and operates without ownership or management linkages to the networks themselves.[160]

The Darknet, largely based on peer-to-peer architecture, offers alternatives to commercial networks. It separates content from intellectual property, treating bits as information rather than as rights-bearing artifacts. It is operated anonymously, without centralized databases or hierarchical client-server architectures. Due to its ungoverned, ad hoc character, legal responsibility for its operation and activities can be pinned to individuals or groups only with difficulty. Napster's legal vulnerability was its centralized index server; as of this writing, enabling a decentralized and noncommercial Darknet is not itself illegal. The Darknet's main challenge to trusted computing is its effectiveness as a meta-distribution platform—a software environment that spreads distribution across all participating nodes through a broad base of computing and networking power. And, despite formally separating the Darknet from rights-bearing artifacts, the record companies link its activities to a black hole of lost revenues, deploring the "anonymizing routers, overseas routers, object fragmentation, or . . . other means to complicate the effort required by law-enforcement to determine the original source of the copyrighted bits."[161] While Biddle et al. do not recommend an Internet policing strategy, they argue that universities and schools should be held partially liable for illicit student activities, that ISPs should be prepared to divulge user identities, that offshore servers should be policed through surveillance of backbone routers, and that local servers connected to offshore servers should be served with individual legal challenges.

However, after reciting this litany of malfeasance, Biddle and colleagues turn the tables. They point out that the Darknet forces businesses to compete "on the Darknet's own terms," by offering "convenience and low cost rather than additional security."[162] The authors thus implicitly recognize that the Darknet is a principal source of pressure on the development of the Celestial Jukebox, acculturating consumers to freely sharing media content while at the same time potentially limiting prices that might otherwise be inflated by collusion between corporations. Moreover, the Darknet promotes open, rather than closed, standards for platforms, and its anonymity encourages users to shun digital-rights management and other trappings of the Celestial Jukebox. By concluding that the Darknet can be overcome only by offering convenience and low cost to consumers, the authors directly challenge the business strategies of their employers.

A more innovative and imaginative industry might find ways to benefit from these Darknet practices. It might, for example, derive market information from real-time aggregate readings of peer-to-peer traffic on the most popular networks. Software for gathering such information is already available from BigChampagne.[163] Such information might enable the industry to modulate the supply of music and develop effective marketing campaigns; in combination with sales data, it might help marketers cross-promote titles, push specific titles, and offer pricing suggestions to retail and online outlets. A pay peer-to-peer system such as the one described in chapter 4, supported by such real-time market research, might sustain the industry even if physical recordings were to disappear entirely.

Meanwhile, the second-generation Internet, still operating among an international consortium of research universities, is beginning to produce disruptive technologies such as LoDN, which provides access to distributed storage space at speeds up to six gigabits per second for applications that need delivery of data or media in real time. These decentralized networks could be permitted to compete with expensive commercial applications that use client-server or "three-tier" architectures, or they could be found illegal or legislated out of existence. Digital art, academic research, cyber countercultures, gift exchanges, and online collaborations generate activities that, directly or indirectly, challenge the controls imposed by the Celestial Jukebox. These challenges have been formalized in new "hacktivist" technologies, but also in intellectual-property-law reform campaigns coordinated by groups like the Center for Digital Democracy, the Media Access Project, and the Electronic Frontiers Foundation. Should the industry's legal efforts prevail, enormous swaths of intellectual production and speech—commercial and noncommercial—could fall under new federal and international criminal sanctions. As chapter 4 illustrates, fundamental values of fair use, privacy, and free speech are at stake.

NOTES

1. Wilfred Dolfsma, "How Will the Music Industry Weather the Globalization Storm?" *First Monday* 5, no. 5 (May 2000), http://firstmonday.org/issues/issue5_5/dolfsma/index.html.

2. Saskia Sassen, "Digital Networks and Power," in *Spaces of Culture: City—Nation—World*, ed. Mike Featherstone and Scott Lash (Thousand Oaks, CA: Sage, 1999).

3. Brian Winston, *Media Technology and Society: A History from the Telegraph to the Internet* (New York: Routledge, 1998).

4. Tom Standage, *The Victorian Internet: The Remarkable Story of the Telegraph and the Nineteenth Century's Online Pioneers* (New York: Walker, 1998).

5. Dan Bricklin, "The Cornucopia of the Commons," in *Peer-to-Peer: Harnessing the Benefits of a Disruptive Technology*, ed. Andy Oram, 59–63 (Cambridge: O'Reilly & Associates, 2001), 63.

6. Cory Doctorow, quoted in Howard Rheingold, *Smart Mobs: The Next Social Revolution* (Cambridge, MA: Perseus Publishing, 2002).

7. Jon Cooper and Daniel M. Harrison, "The Social Organization of Audio Piracy on the Internet," *Media Culture & Society* 23 (2001): 73.

8. Reebee Garofalo, "From Music Publishing to MP3: Music and Industry in the Twentieth Century," *American Music* 17, no. 3 (1999): 349.

9. Cooper and Harrison, "Music and Industry."

10. Andrew Leyshon, "Scary Monsters? Software Formats, Peer-to-Peer Networks, and the Spectre of the Gift," *Environment and Planning D: Society and Space* 21 (2003).

11. Colin Levy, "New Technology Calls the Tunes," *Wall Street Journal*, 8 Mar. 1999, A18.

12. Lee Gomes, "Free Tunes for Everyone!" *Wall Street Journal*, 15 June 1999, B1.

13. Matt Richtel, "Music Industry Loses a Bid to Stop Internet Recording," *New York Times*, 28 Oct. 1998, C9.

14. Nelson Minar and Marc Hedlund, "A Network of Peers: Peer-to-Peer Models through the History of the Internet," in *Peer-to-Peer*, ed. Oram, 5.

15. Minar and Hedlund, "Network of Peers," 7.

16. Clay Shirky, "Listening to Napster," in *Peer-to-Peer*, ed. Oram, 21–22.

17. Milton Mueller, *Ruling the Root: Internet Governance and the Taming of Cyberspace* (Boston: MIT Press, 2004).

18. Shirky, "Listening," 25.

19. Philip Hayward, "Enterprise on the New Frontier: Music, Industry and the Internet," *Convergence* 1, no. 2 (1995): 29–44.

20. Niels Schaumann, "Copyright Infringement and Peer-to-Peer Technology," *William Mitchell Law Review* 28, no. 3 (2002): 1034.

21. Quoted in Schaumann, "Copyright," 1033–34 (emphasis in original).

22. Leyshon, "Scary Monsters?"

23. John Alderman, *Sonic Boom: Napster, P2P, and the Battle for the Future of Music* (Cambridge, MA: Perseus Press, 2001).

24. Schaumann, "Copyright," 1036.

25. Schaumann, "Copyright," 1038.

26. Vincent Mosco, *The Pay-Per Society: Computers and Communication in the Information Age* (Mahwah, NJ: Ablex, 1990).

27. Keith Negus, *Music Genres and Corporate Cultures* (London: Routledge, 1999), 56–58.

28. Negus, *Music*, 55.

29. Julian Stallabrass, *Gargantua: Manufactured Mass Culture* (New York: Verso, 1996), 62–63.

30. Quoted in Nick Paton, "Mom, I Blew Up the Music Industry," *The Observer*, 21 May 2000, http://www.observer.co.uk/review/story/0,6903,223075,00.html.

31. Martin Peers, "In the Groove," *Wall Street Journal*, 20 Mar. 2000, R14.

32. Minar and Hedlund, "Network of Peers," 4–5.

33. Anna Wilde Mathews and Bruce Orwall, "Record Firms, Studios Sue Scour, Alleging Theft via Its Site on Web," *Wall Street Journal*, 21 July 2000, B6.

34. Matt Richtel, "Music and Movies Web Site in Bankruptcy-Law Filing," *New York Times*, 14 Oct. 2000, B4.

35. Ben Charny, "The New Scour: Users Will Pay," *ZDnet.com*, 14 Dec. 2000, http://news.zdnet.com/2100-9595_22-526419.html.

36. Beth L. Krigel, "MP3.com Flies in Trading Debut," *CNet News*, 21 July 1999, http://news.com.com/2100-1023-228821.html?legacy=cnet.

37. Sara Robinson, "MP3.com Plans to Let Users Store Music Files on Its Site," *New York Times*, 12 Jan. 2000, C2.

38. Leyshon, "Scary Monsters," 15.

39. Amy Harmon with John Sullivan, "Music Industry Wins Ruling in U.S. Court." *New York Times*, 29 Apr. 2000, B1.

40. Matt Richtel, "Two Record Labels Settle Copyright Suit with MP3.com," *Wall Street Journal*, 10 June 2000, B1.

41. Matt Richtel, "Chief of MP3.com Testifies in Music Copyright Hearing," *New York Times*, 29 Aug. 2000, C6.

42. Amy Harmon, "Judge Rules Against MP3 on CD Copying," *New York Times*, 7 Sept. 2000, C1.

43. Anna Wilde Mathews, "MP3.com May Be Playing Their Songs after Publishers Agree to Tentative Deal," *Wall Street Journal*, 19 Oct. 2000, B20; Anna Wilde Mathews and Colleen DeBaise, "Judge Says MP3.com Broke Laws, Owes Damages to Seagram Unit," *Wall Street Journal*, 7 Sept. 2000, B16.

44. Amy Harmon, "Deal Settles Suit against MP3.com," *New York Times*, 15 Nov. 2000, C1.

45. Mark Lewis, "UMG Fear or MP3.com Bankruptcy? Inside the Settlement," *Webnoize*, 15 Nov. 2000, http://www.news.webnoize.com/item.rs?ID-11067.

46. Charles Mann, "What Goes Around Comes Around Dept.: Song Publishers Sue Universal Music for Copyright Infringement," *Inside.com*, 8 Dec. 2000, http://www.inside.com/jcs/Story?articleid=17380&pod_id=9.

47. Matt Carolan, "It's a Shame about MP3.com," *Zdnet*, 12 Sept. 2000, http://cma.zdnet.com/texis/techinfobase/techinfobase/+rwo_qr+_+sWKs/zdisplay.html.

48. Gary Gentile, "Randy Newman, Tom Waits Sue MP3.com," *Salon*, 8 May 2001, http://news.com.com/2110-1023-257241.html?legacy=cnet.

49. Amy Harmon, "MP3.com to Restart Its Music Service, for Those Willing to Pay," *New York Times*, 5 Dec. 2000, C4.

50. Andrew Sorkin, "Vivendi in Deal for MP3.com to Lift Online Distribution," *New York Times*, 21 May 2001, C1.

51. Gwendolyn Mariano, "MP3.com Splits into Two," *CNet News*, 19 Oct. 2001, http://news.cnet.com/news/0-1005-200-7586213.html?tag=mn_hd.

52. Andrew Orlowski, "Vivendi Spinoff Takes MP3.com Archive Private," *The Register*, 1 Dec. 2004.

53. Odd Arild Skaflestad and Nina Kaurel, "Peer-to-Peer Networking: Configuring Issues and Distributed Processing" (paper presented at NTNU conference, 2001), 4, http://www.item.ntnu.no/fag/SIE50AC/P2P.pdf.

54. Media Metrix, "Press Release," 11 Oct. 2000, 1, http://www.mediametrix.com/press/releases/20001005.jsp?language=us.

55. Susan Stellin, "Napster Use Quadrupled in Five Months," *New York Times*, 12 Sept. 2000, C6.

56. *The Hollywood Reporter*, "Study: File Sharing Cutting into Retail," 24 May 2000.

57. Renee Graham, "My Torrid Love Affair with Napster," *Boston Globe*, 13 Mar. 2001, http://msl1.mit.edu/ESD10/bglobe_mar_13_01_d.pdf.

58. Jon Pareles, "MP3.com Hopes for Deal in Copyright Suit," *New York Times*, 1 May 2000, C6.

59. Matt Richtel and Neil Strauss, "Metallica to Try to Prevent Fans from Downloading Recordings," *New York Times*, 3 May 2000, C1.

60. Ryan Tate, "Record Industry Argues—Again—for Napster Injunction," *Upside*, 11 Sept. 2000, http://web.archive.org/web/20001017123744/http://www.upside.com/News/39bd3c270.html.

61. John Heilemann, "David Boies: The Wired Interview," *Wired*, Oct. 2000, http://www.wired.com/wired/archive/8.10/boies.html.

62. Lee Gomes, "Napster Is Ordered to Stop the Music," *Wall Street Journal*, 27 July 2000, A3.

63. Lee Gomes, "1984 Sony Case Key to Napster Legal Strategy," *Wall Street Journal*, 13 Sept. 2000, B1.

64. Heilemann, "Boies."

65. Lee Gomes, "Napster Stakes Out 'Fair Use' Defense on Music Sharing," *Wall Street Journal*, 5 July 2000, B2.

66. Alex Salkever, "Napster's Battle," *Businessweek Online*, 19 July 2000, http://www.businessweek.com/bwdaily/dnflash/july2000/nf00719a.htm.

67. Lee Gomes, "Napster Is Ordered to Stop the Music," *Wall Street Journal*, 27 Sept. 2000, A3.

68. Brad King, "New School of Thought on Piracy," *Wired*, 9 June 2000, http://www.wired.com/news/culture/0,1284,36875,00.html.

69. Brian Ploskina, "Numbers Rock 'n' Roll in Napster Dispute," *Inter@ctive Week*, 19 June 2000, http://wwwpub.utdallas.edu/~liebowit/knowledge_goods/ploskina.pdf.

70. Alex Berenson and Matt Richtel, "Heartbreakers, Dream Makers," *New York Times*, 25 June 2000, C1.

71. Jim Hu and John Borland, "Can Napster Survive the Bertelsmann Deal?" *CNet News*, 1 Nov. 2000, http://news.com.com/2100-1023-247956.html?legacy=cnet.

72. Lee Gomes et al., "Bertelsmann, Napster Agree on Service," *Wall Street Journal*, 1 Nov. 2000, A3.

73. Brad King, "Napster Secures New Format," *Wired News*, 2 Mar. 2001, http://www.wired.com/news/print/0,1294,42097,00.html.

74. Matt Richtel, "Internet Service to Charge a Fee for Music Rights," *New York Times*, 1 Nov. 2000, A1; Gomes et al., "Bertelsmann."

75. Charles Mann, "The Secret's Out: Napster and Bertelsmann Finally Reveal Blueprint for New Version of File-Swapping Service," *Inside.com*, 16 Feb. 2001, http://www.inside.com/jcs/Story?article_id=23794&pod_id=9.

76. Todd Spangler, "The Napster Mirage," *Inter@ctive Week*, 24 July 2000, http://www.zdnet.com/intweek/stories/news/0,4164,2607261,00.html.

77. Lee Gomes and Anna Wilde Mathews, "Napster Suffers a Rout in Appeals Court," *Wall Street Journal*, 13 Feb. 2001, A3; also see Matt Richtel, "Appellate Court Backs Limitation on Copying Music," *New York Times*, 13 Feb. 2001, A1.

78. Charles Mann, "The Day After," *Inside.com*, 13 Feb. 2001, http://detritus.net/contact/rumori/200103/0307.html.

79. Matt Richtel, "Judge Orders Napster to Police Trading," *New York Times*, 7 Mar. 2001, C1.

80. Don Clark, "Napster Offers Annual Fees to CD Labels," *Wall Street Journal*, 21 Feb. 2001, B6.

81. Brad King, "Labels to Napster: Download This," *Wired.com*, 22 Feb. 2001, http://www.wired.com/news/business/0,1367,41941,00.html.

82. Brad King, "RIAA Head: Napster is Done," *Wired.com*, 2 May 2001, http://www.wired.com/news/mp3/0,1285,43487,00.html.

83. Sue Zeidler, "Report: Napster Users Don't Share Well with Others," Reuters, 27 June 2001, http://www.idobi.com/news.wml?200106287.

84. John Borland, "Napster's Tab: $100 Million and Climbing," *CNet News*, 26 Sept. 2001, http://news.cnet.com/news/0-1005-202-7310893.html.

85. Matt Richtel, "Napster Wins One in Music Case," *New York Times*, 23 Feb. 2002, B1.

86. Matt Richtel, "Napster Wins One."

87. Brad King, "Judge: If You Own Music, Prove It," *Wired.com*, 22 Feb. 2002, http://www.wired.com/news/print/0,1294,50625,00.html.

88. Nick Wingfield, "Napster Files for Chapter 11 Shelter," *Wall Street Journal*, 4 June 2002, B6.

89. Matt Richtel, "Napster Says It Is Likely to Be Liquidated," *New York Times*, 4 Sept. 2002, C2; Phyllis Furman, "Napster's Back in Town: Roxio Pays 5M for Music-Swapper," *New York Daily News*, 16 Nov. 2002, 17.

90. Furman, "Napster's Back"; Nick Wingfield, "Roxio Agrees to Acquire Napster Assets," *Wall Street Journal*, 18 Nov. 2002, B4.

91. Furman, "Napster's Back," 17.

92. S. Colella, "Returning Students Required to Register for Napster," *The Digital Collegian* (Pennsylvania State University), 10 Sept. 2004, http://www.collegian.psu.edu/archive/2004/09/09-10-04tdc/09-10-04dnews-06.asp.

93. OpenNap, http://opennap.sourceforge.net/.

94. Matt Richtel, "With Napster Down, Its Audience Fans Out," *New York Times*, 20 July 2001, A1.

95. Colin Beavan, "Aimster: A Deft and Possibly Lethal Joining of Napster and AOL," *MSNBC.com*, 12 Dec. 2000, http://www.msnbc.com/news/502033.asp?cpl=1; King, "Napster Secures."

96. Matt Richtel, "New Suit Filed to Bar Trading Music on Net," *New York Times*, 25 May 2001, C2.

97. United States Court of Appeals for the Seventh Circuit, No. 02-4125 IN RE: AIMSTER COPYRIGHT LITIGATION. APPEAL OF: JOHN DEEP, Defendant. Appeal from the United States District Court for the Northern District of Illinois, Eastern Division. No. 01 C 8933-Marvin E. Aspen, Judge, ARGUED JUNE 4, 2003, DECIDED JUNE 30, 2003, http://www.riaa.com/news/newsletter/pdf/aimster20030630.pdf.

98. John Markoff, "The Concept of Copyright Fights for Internet Survival," *New York Times*, 10 May 2000, 1.

99. Steven Levy, "The Noisy War over Napster," *Newsweek*, 5 June 2000.

100. Jon Orwant, "What's on Freenet?" *The O'Reilly Network*, 21 Nov. 2000, http://www.oreillynet.com/pub/a/p2p/2000/11/21/freenetcontent.html.

101. Lee Gomes, "Gnutella, New Music-Sharing Software, Rattles the CD Industry," *Wall Street Journal*, 4 May 2000, B10.

102. David Streitfeld, "The Web's Next Step: Unraveling Itself," *Washington Post*, 18 July 2000, A1.

103. Janelle Brown, "The Gnutella Paradox," *Salon*, 29 Sept. 2000, http://www.salon.com/tech/feature/2000/09/29/gnutella_paradox/index.html.

104. Skaflestad and Kaurel, "Peer-to-Peer Networking," 6.

105. John Markoff, "Many Take, But Few Give on Gnutella," *New York Times*, 21 Aug. 2000, C4; Brown, "The Gnutella Paradox."

106. Andy Oram, "Peer-to-Peer Makes the Internet Interesting Again," *The O'Reilly Network*, 22 Sept. 2000, http://linux.oreillynet.com/lpt/a/linux/2000/09/22/p2psunnit.html.

107. Mark Lewis, "Does Morpheus' Architecture Save MusicCity from Legal Liability?" *Webnoize*, 23 Aug. 2001, http://www.dtype.org/pipermail/p2p-legal/2001-August/000041.html.

108. Kevin Delaney, "KaZaA Founder Peddles Software to Speed File Sharing," *Wall Street Journal*, 8 Sept. 2003, B1.

109. Amy Harmon, "Music Industry in Global Fight on Web Copies," *New York Times*, 7 Oct. 2001, A1.

110. Matt Richtel, "In Free Music Software, a Hidden Fee-Based Service," *New York Times*, 3 Apr. 2002, C2.

111. Todd Woody, "The Race to Kill KaZaA," *Wired.com*, Feb. 2003, http://www.wired.com/wired/archive/11.02/kazaa_pr.html.

112. Woody, "The Race."

113. John Borland, "Judge: File-Swapping Tools Are Legal," *CNet News*, 25 Apr. 2003, http://news.com.com/2100-1027-998363.html.

114. Ben Fritz, "Peering at Piracy? Court Grapples Grokster-Streamcast Case," *Variety*, 3 Feb. 2004, http://www.variety.com/article/VR1117899510?categoryid=18&cs=1.

115. Saul Hansell, "Aiming at Pornography to Hit Music Piracy," *New York Times*, 7 Sept. 2003, A1.

116. Nick Wingfield, "Behind the Fake Music," *Wall Street Journal*, 11 July 2002, D1.

117. Guy Dixon, "Dividing the Spoils," *Globe and Mail.com*, 29 July 2003, http://www.theglobeandmail.com/servlet/ArticleNews/TPPrint/LAC/20030729/PIRACY29/TPEntertainment/.

118. Ted Bridis, "Hatch Takes Aim at Illegal Downloading," *Washington Post*, 17 June 2003, http://www.washingtonpost.com/ac2/wp-dyn/A6241-2003Jun17.

119. U.S. Court of Appeals, "In Re: Aimster," 4.

120. Jeff Leeds, "The Labels Strike Back," *Los Angeles Times*, 10 Sept. 2003, C1.

121. Amy Harmon, "Suit Settled for Students Downloading Music Online," *New York Times*, 2 May 2003, A22.

122. Amy Harmon with John Sullivan, "Music Industry Wins Ruling in U.S. Court," *New York Times*, 29 Apr. 2000, B1.

123. Jon Healey, "Record Companies to Offer Amnesty to File Sharers, with Conditions," *Los Angeles Times*, 5 Sept. 2003, C1.

124. Lorena Mongelli, "Music Pirate," *The New York Post*, 9 Sept. 2003; Rick Munarriz, "Steal This Column," *Motley Fool*, 12 Sept. 2003, http://www.fool.com/news/commentary/2003/commentary030912ram.htm.

125. Nick Wingfield and Nick Baker, "RIAA Targets Are Surprised by Piracy Suits," *Wall Street Journal*, 10 Sept. 2003, B1.

126. Benny Evangelista, "Download Lawsuit Dismissed," *San Francisco Chronicle*, 25 Sept. 2003, http://www.sfgate.com/cgi-bin/article.cgi?f=/c/a/2003/09/25/BUGJC1TO2D1.DTL.

127. John Schwartz, "She Says She's No Music Pirate. No Snoop Fan, Either," *New York Times*, 25 Sept. 2003, C1.

128. Munarriz, "Steal This Column."

129. Katie Dean, "Marking File Traders as Felons," *Wired.com*, 2003 http://www.wired.com/news/business/0,1367,58081,00.html.

130. Ethan Smith, "Music Industry's Move against Swappers Hits a Snag Just as Impact Takes Hold," *Wall Street Journal*, 22 Dec. 2003, B1.

131. Alex Salkever, "Big Music's Worst Move Yet," *BusinessWeek Online*, 27 Jan. 2004, http://www.businessweek.com/print/technology/content/jan2004/tc20040127_2819_tc047.htm?tc.

132. Peter Shinkle, "Some Accused of Music Downloading Turn to Bankruptcy," *St. Louis Post-Dispatch*, 26 Sept. 2004, http://www.stltoday.com/stltoday/news/stories.nsf/stlouiscitycounty/story/E4C3B28FB081CED286256F1C00177CE3?OpenDocument&Headline=Some+accused+of+muic+downloading+turn+to+bankruptcy; Andy Sullivan, "Record Industry Sues 762 for Net Music Swaps," *USA Today*, 10 Sept. 2004, http://www.usatoday.com/tech/news/techpolicy/2004-09-30-riaa-suit_x.htm.

133. Lee Rainie et al., "Pew Internet Project and comScore Media Metrix Data Memo," *Pew Internet Project*, Apr. 2004, http://www.pewinternet.org/PPF/r/124/report_display.asp.

134. Jefferson Graham, "Online File Swapping Endures: Users Undaunted by Threat of Suits," *USA Today*, 12 July 2004, A1.

135. Seth Schiesel, "File Sharing's New Face," *New York Times*, 12 Feb. 2004, G1.

136. Alec Foege, "Record Labels Are Hearing an Angry Song," *New York Times*, 11 June 2000, BU4.

137. Tony Smith, "Music Biz Changes Tune on Net Threat," *TheRegister*, 2 June 2000, http://www.theregister.co.uk/2000/06/02/music_biz_changes_tune/.

138. Jane Black, "Big Music's Broken Record," *BusinessWeek Online*, 13 Feb. 2003, http://www.businessweek.com/print/technology/content/feb2003/tc20030213_9095_tc078.htm?tc.

139. *Yahoo Daily News*, "Napster, Shmapster: 'N Sync Rules Record Y2K," 5 Jan. 2001, http://dailynews.yahoo.com/h/eo/20010105/en/napster_shmapster_n_sync_rules_record_y2k_3.html.

140. Eric Hellweg, "Show Me the (Missing) Money," *Business2.com*, 3 Jan. 2001, http://www.business2.com.

141. *The Economist*, "Music's Brighter Future," *Economist.com*, 28 Oct. 2004, http://www.economist.com/business/displayStory.cfm?story_id=3329169.

142. Dan DeLuca, "Rock of (Middle) Ages," *Philadelphia Enquirer*, 23 Apr. 2002, E1.

143. Charles Goldsmith, "Global CD Sales Last Year Declined for the First Time," *Wall Street Journal*, 17 Apr. 2002, B2.

144. Charles Mann, "Napster's Billion-Dollar Pledge," *Inside.com*, 23 Feb. 2001, http://www.inside.com/jcs/Story?article_id=24235&pod_id=9.

145. Anna Wilde Mathews, "Music Samplers on Web Buy CDs in Stores," *Wall Street Journal*, 15 June 2000, A3.

146. Mathews, "Music Samplers," A3; Mark Latonero, "Survey of MP3 Usage: Report on a University Consumption Community," Annenberg School of Communication, University of Southern California, June 2000.

147. Matt Richtel, "Access to Free Online Music Is Seen as a Boost to Sales," *New York Times*, 6 May 2002, C6.

148. John Schwartz, "A Heretical View of File Sharing," *New York Times*, 5 Apr. 2004, C1, and http://www.unc.edu/~cigar/papers/FileSharing_March2004.pdf.

149. Sean Silverthorne, "Music Downloads: Pirates—or Customers?" *Harvard Business School Working Knowledge*, 21 June 2004, http://hbswk.hbs.edu/tools/print_item.jhtml?id=4206&t=leadership.

150. Silverthorne, "Music Downloads."

151. Cary H. Sherman, RIAA, letter to Senator Norm Coleman, 14 Aug. 2003, http://www.eff.org/IP/P2P/RIAA-coleman.pdf.

152. Eliot Van Buskirk, "Allow Me to Induce Myself," *CNet*, 7 July 2004, http://reviews.cnet.com/4520-6450_7-5142741-1.html.

153. Grokster and EFF [Electronic Frontiers Foundation], (2004, Nov. 8). Brief in opposition. Metro-Goldwyn-Mayer Studios Inc., et al., Petitioners, v. Grokster, Ltd., et al., Respondents. On petition for writ of certiorari to the United States Court of Appeals for the Ninth Circuit, http://www.eff.org/IP/P2P/MGM_v_Grokster/20041108_Final_Brief.pdf.

154. Alderman, *Sonic Boom*.

155. N'Gai Croal, "Sony's Digital Dilemma," *Newsweek*, 6 Jan. 2001, http://www.msnbc.com/news/512640.asp.

156. Thomas E. Weber, "'Peer to Peer' Connections Make a Smarter Internet," *Wall Street Journal*, 11 Nov. 2000.

157. Bill Rosenblatt and Gail Dykstra, "Integrating Content Management with Digital Rights Management: Imperatives and Opportunities for Digital Content Lifecycles," *Giantsteps Media Technology Strategies*, 14 May 2003 (italics in original), http://www.drmwatch.com/resources/whitepapers/article.php/3112011.

158. Seth Schoen, "Trusted Computing: Promise and Risk," *EFF.org*, http://www.eff.org/Infrastructure/trusted_computing/20031001_tc.php.

159. Peter Biddle, Paul England, Marcus Peinado, and Bryan Willman, "The Darknet and the Future of Content Distribution" (paper presented at the 2002 ACM Workshop on Digital Rights Management, Washington, D.C., 18 Nov. 2002), http://crypto.stanford.edu/DRM2002/darknet5.doc.

160. Biddle et al., "The Darknet," 1.

161. Biddle et al., "The Darknet," 7.

162. Biddle et al., "The Darknet," 16.

163. Alex Veiga, "Music Labels Tap Downloading Networks," Associated Press, 7 Nov. 2003, http://story.news.yahoo.com/news?tmpl=story&u=/ap/file_swapping_intelligence.

4

The Jukebox Implemented

The Internet was long dismissed as a competitive threat to the entertainment business. Citing a number of corporate leaders, Robert McChesney predicted that Internet media distribution was unlikely in the short or medium term, given a torpid industry adoption of e-business models and a slow rollout of last-mile broadband infrastructure:

> It will be many years before the Internet can possibly stake a claim to replace television as the dominant medium in the United States. . . . Rupert Murdoch, whose News Corporation has been the most aggressive of the media giants of cyberspace, states that establishing an information highway "is going to take longer than people think." He projects that it will take until at least 2010 or 2015 for a broadband network to reach fruition in the United States and western Europe, and until the middle of the 21st century for it to begin to dominate elsewhere. Even Bill Gates . . . acknowledges that the Internet as a mass medium "is going to come very slowly." . . . As MCA president Frank Biondi put it in 1996, media firms "don't even think of the Internet as competition."[1]

These sentiments may seem to us today to result from a lack of vision, but they reflect the limitations of the legal, software, and telecommunication infrastructures of their time. As recently as 2000, online distribution was not considered a significant factor in studies of the music business.[2] However, the falling price of multimedia PCs and the increased access to broadband delivery in homes and offices changed this attitude very quickly. Napster caught the industry's attention as a viable system for space shifting and time shifting music collections, and its rapid adoption took seasoned media critics by surprise—and engulfed the music industry in fear.

And this was not without reason. For years, the music industry made profits from selling physical recordings through retailers; in the future, many observers envision the market-based economy of buyers and sellers of hard goods replaced by a network-based economy of servers and clients of flows of assets. Rifkin argues that

> we increasingly pay for the time we use things, rather than for the things themselves. . . . By the middle of the century, property exchange in markets will have disappeared from much of commercial life, and been replaced by up-to-the-moment access relationships inside vast commercial networks. . . . Users will pay to experience the music, not own it in the form of a physical product.[3]

Though recorded artifacts doubtlessly will persist in music markets for some time, the industry has sought to retool for the Internet. In the hope of luring back file traders and deterring artists from bypassing record companies, the industry first tried to construct an Internet distribution system modeled on broadcast, with streaming channels and personalized playlists in addition to the pay-per-burn services. They quickly abandoned these music-service providers for third-party distribution sites, or clearinghouses. Although the recording industry's online efforts remain in flux, to remain viable it must develop a business model that requires consumers to "pay to experience the music" through systems based on trusted computing.

As described in chapter 3, the recording industry could have worked peer-to-peer systems to its advantage. Yet its reflexive response to disruption and its desire for control at all costs (to the extent of suing its own customers) have led the industry instead to favor a technological lockdown of Web "assets," including music and video files, software, digital photographs, and e-books. And it has set in place a legal infrastructure that makes this possible: the 1996 Telecommunications Act, the Digital Millennium Copyright Act (DMCA), and the Sonny Bono Copyright Term Extension Act, the latter two passed in 1998. These laws set terms for the punishment of copyright outlaws in the United States that extend internationally through the World Intellectual Property Organization, a division of the World Trade Organization.[4]

Since the passage of these laws and the victory over Napster, the recording industry has sought to move its digital content to the Internet through a system analogous to pay-per-view TV. It seeks to protect its control by developing lockbox technology, by lobbying and litigating to stem the development of technologies that enable resistance, and by "reeducating" consumers through propaganda campaigns and subjecting dissidents to disciplinary legal actions. Moreover, it hopes simultaneously to develop and refine the narrowcasting strategies of radio and cable television through personalized "stations," consumer-provided playlists, and pay-per downloads, catering to individual consumer tastes by knowing what the customer wants before the customer knows or wants it.

The Big Five's first strategy for achieving all this was through exclusive ownership of the digital-music spigots, which would allow them to bypass traditional promotional media (print, radio, and MTV) and physical markets (retailers) for direct, real-time access to customers. In December 2001, the Big Five unveiled their first music-service providers, or MSPs: MusicNet and Pressplay. The latter was the creation of Universal and Sony, the former of Warner, BMG, and EMI. They were soon joined by eMusic, in which Universal also participated. While each MSP featured the partial catalogs of its owners, each refrained from offering catalogs to its rival services. Because music-CD releases were not cross-licensed between record labels and constituted the bulk of the catalogs for the MSPs, the results for each were "akin to a radio that gets only half the stations."[5] Subsequent cross-licensing arrangements between the Big Five attempted to redress this problem, but cross-licensing also led to an oligopoly for online distribution. Except for limited offerings to Rhapsody (which was sold by Listen.com to RealNetworks), the Big Five refused to license their catalogs to competing Internet start-ups such as Musicmaker.com and Riffage.com.

This blatant power grab created anxiety in an otherwise agreeable and acquiescent U.S. legislature, and the subsequent demise of these start-ups prompted two U.S. legislators to propose a compulsory licensing bill, the Music Online Competition Act (MOCA) to "require the major music companies involved in the joint ventures to make the same terms available to independent online music distributors."[6] A co-sponsor of MOCA said the Big Five "want to be the owner of the content, the producer, the wholesaler, and the record store online. And with cross-licensing agreements, they are in a position to leverage their collective market power to the exclusion of the competition."[7]

The licensing bottleneck also attracted the attention of the U.S. Department of Justice, which initiated an investigation into its anticompetitive effects. After being hounded by the chair of the Senate Judiciary Committee, Orrin Hatch, the Big Five finally responded with a flurry of cross-licensing deals to other services.[8] The Department of Justice closed its investigation in late 2003, claiming that

> consumers now have available to them an increasing variety of authorized outlets from which they can purchase digital music, and consumers are using those services in growing numbers. . . . The Division found no impermissible coordination among the record labels as to the terms on which they would individually license their music to third-party services.[9]

And so the legal threat to the Celestial Jukebox by antitrust regulators was averted early. However, almost immediately after the creation of these services, most of the majors divested from them. Sony Music and Universal Music Group exited from Pressplay/Napster, and Universal from eMusic, leaving MusicNet as the lone MSP. The companies turned instead to music clearinghouses—third parties, such as iTunes and the Microsoft music store, that broker catalog content from the Big Four but are owned and operated independently of them (table 4.1 lists the major online distributors). By 2005, the iTunes store captured 70 percent of all music distributed legally through the Internet.[10]

Why did the Big Four abandon the MSPs? Pressures for short-term profits, coupled with the economic risks posed by developing an online music infrastructure, convinced the Big Four that they stood to gain more by licensing their primary economic assets (recordings) out to clearinghouses. These clearinghouses would then sort out infrastructural problems, thus insulating the Big Four from research-and-development costs. By the time of their divestiture, Universal and Sony had each poured $30 million into Pressplay and had attracted fewer than 50,000 subscribers for their trouble. Vivendi Universal sponsored the eMusic MSP until their divestiture in 2003; eMusic then relaunched itself as a clearinghouse. Since each of the Big Four receive about thirty cents from every ninety-nine-

Table 4.1. Leading MSPs and Clearinghouses, 2004

Service	*Subscription and/or Download*	*Pricing*	*Restrictions on Copying*	*File Format*	*DRM System*
Connect (Sony)	Download	$0.99/track	Playlists of purchased tracks can be burned up to seven times.	ATRAC	Sony OpenMG
eMusic	Download	$9.99/month for forty downloaded tracks	None	MP3	None
iTunes	Download	$0.99/track	Playlists of purchased tracks can be burned up to seven times.	AAC	Fair Play
MSN (Microsoft)	Download	$0.99/track	Playlists of purchased tracks can be burned up to seven times.	WMA	Microsoft
MusicMatch (Yahoo!)	Download and subscription	$4.99/month for MusicMatch on Demand; $0.99/track for downloads	Playlists of purchased tracks can be burned up to seven times.	WMA	Microsoft
Napster (formerly Pressplay)	Download and subscription	$9.95/month for streaming; $0.99/track for downloads; $14.95/ month for To Go service (unlimited transfers to players)	Playlists of purchased tracks can be burned up to seven times.	WMA	Microsoft
Rhapsody/Real Player	Download and subscription	$9.95/month (Rhapsody); $0.99/track for downloads	Playlists of purchased tracks can be burned up to five times.	WMA	Microsoft
Wal-Mart	Download	$0.88/track	Individual purchased tracks can be burned up to ten times	WMA	Microsoft

Source: Companies' marketing and promotional websites.

cent download from a clearinghouse, the Big Four calculated that it was better to sell to every possible clearinghouse than to focus on running their own operations. The Big Four and their parent companies are keenly interested in the hardware and software possibilities of digital delivery, since the recording industry's developing online model will provide a template for the delivery of content such as video, film, and text. In clearinghouses as well as MSPs, the Big Four were careful as always to introduce intellectual-property bottlenecks at each link in the chain. Their care, taken early, protects the seeds of what could become a bountiful field of revenue in the near future. Figure 4.1 projects online music-sales growth for U.S. and global markets.

Figure 4.1. Online Music Sales, 2000–2007

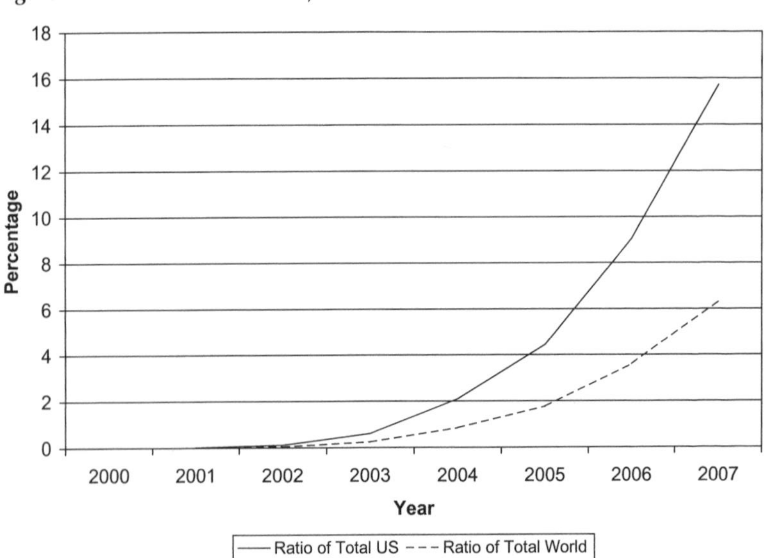

Note: Includes online music subscriptions, album downloads, and single song downloads; excludes sales of tangible formats.
Sources: "Metrics," *EContent* 25, no. 11 (15 Nov. 2002); "Reports Predict That the Value of Global Recorded Music Sales Will Be Lower in 2005 Than in 2001," *Music & Copyright* 1 (11 Dec. 2002); "RIAA Cites Downloads as Factor in Declining Sales," *TWICE* 17, no. 6 (11 Mar. 2002): 37; "Entertainment Revenue Reaches $53 billion in '03," *Entertainment Marketing Letter* 17, no. 3 (1 Feb. 2004): 1; "UMG Choice Music to Industry Ears: Many See VU Decision to Keep Asset as Vote of Confidence for Down Sector (Money)," *Hollywood Reporter* 379, no. 19 (3 July 2003): 6.

Clearly, the market offers great potential for profit from online music sales; total sales figures for online music markets in the United States swelled dramatically in 2002. While the total for 2004 was still small—4 percent—the U.S. online market is projected to quadruple to 16 percent by 2007. Globally, online music sales are growing at a considerably slower rate, with only 1 percent total music sales in 2004, projected at 6 percent by 2007. However, the U.S. market has an increasing proportion of the global market's music sales, and thus, despite the fact that industry ownership has become increasingly transnational, the Celestial Jukebox caters to a largely U.S. audience (see figure 4.2).

In the remainder of this chapter, we will examine the industry's attempts to carefully define and control its online audience. The Celestial Jukebox uses its online distribution platform to conduct a "panoptic sort"[11] in which no transaction—no download, no stream, and no interaction with a media player—goes unaccounted. This

Figure. 4.2. U.S. Music Sales as a Percentage of Global Music Sales, 1999–2003

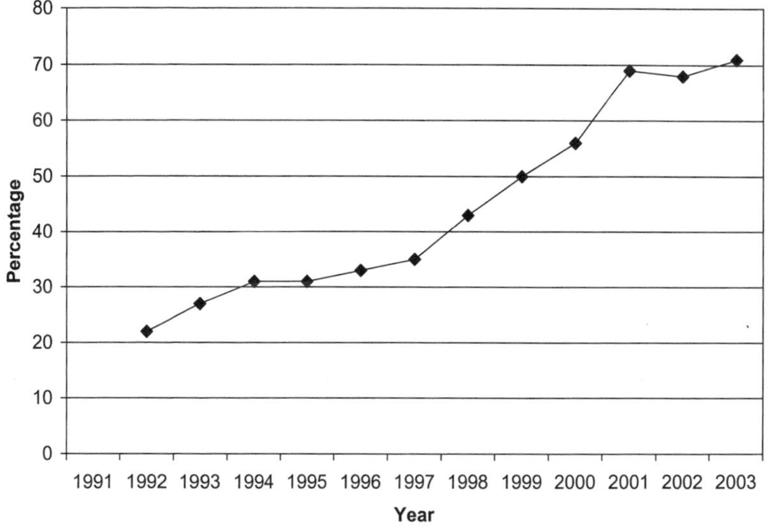

Sources: "Reports Predict That the Value of Global Recorded Music Sales Will Be Lower in 2005 Than in 2001," *Music & Copyright* 1 (11 Dec. 2002); "Napster: Catalyst for a New Industry or Just Another Dot-com? (The Ivey Case Study©): Part 3 of 3," *Ivey Business Journal* 66, no. 3 (Jan. 2002): 45(11); "RIAA Cites Downloads as Factor in Declining Sales," *TWICE* 17, no. 6 (11 Mar. 2002): 37.

"panoptic sort" is made possible through two discrete but intermeshing software applications—customer-relationship management (CRM), which promises to steer the "right" content to the "right" consumers, and digital-rights management (DRM), which enforces restrictions on its use. CRM and DRM may make the Celestial Jukebox commercially viable in the long term. However, their application may also be used to infringe upon consumer rights of privacy, fair use, and free speech. We examine the components, history, and problems of these systems in turn.

CUSTOMER-RELATIONSHIP MANAGEMENT

Customer-relationship management is based on personalization systems, which seek to build brand loyalty by creating an online "experience" tailored to customer preferences. Customers typically access such systems through a "portal" or "My Service" interface, which allows them to customize the information they receive, such as news, messages, recommendations, and billing notices. These systems recognize and track returning customers; the more the customer uses the service, the more accurate its suggestions become. Such systems assemble marketing dossiers as they track interactions and purchases; this information, about both individual and general consumer behavior, can be used to hone in-house marketing efforts and can also be traded between corporate divisions or sold to outside interests. Direct-mail firms and the U.S. Postal Service have used similar software for decades to sort information into databases that can be rented and sold to marketers.

Constructing the "community" central to the Celestial Jukebox's operation requires, paradoxically, that its users be individuated and isolated. In real communities, whether in physical space or cyberspace, members share affinities, interests, and needs; this commonality is recognized and mediated by the members themselves. In contrast, online music stores construct the *appearance* of community while largely denying members the ability to interact directly. And the members have so little control of the terms of their membership that the intimate information they are required to surrender may be transferred without notice to anybody.

Collaborative Filtering and Genre Matching

In the recording industry, the personalization systems responsible for engendering "community" fall into three categories: collaborative filtering, which suggests content based on the user's purchasing history and volunteered comments from the user and others; human-based genre/mood matching, in which experts classify and categorize individual music tracks into logical groupings; and "listening machines," which analyze the actual wave forms of recordings so as to compare their melody, tempo, harmony, timbre, and density.

Collaborative filtering is meant to serve as an automated equivalent to word of mouth. A user's site-navigation patterns, purchasing history, and volunteered feedback are compared to those of other users; recommendations are based on the resulting matches. An early experiment in a collaborative filtering system, a 1992 project of Xerox PARC called Tapestry[12] was followed soon after by the GroupLens project, which tracked each item a user rated along with a score for preference. It then estimated which users were "good predictors" by finding similarities between user tastes and then used these "good predictors" to find new items to recommend. The first music collaborative filtering system, RINGO, was created at the MIT Media Lab in July 1994. One user described the service:

> What RINGO did was simple. It gave you 20-some music titles by name, then asked, one by one, whether you liked it, didn't like it, or knew it at all. That initialized the system with a small DNA of your likes and dislikes. Thereafter, when you asked for a recommendation, the program matched your DNA with that of all the others in the system. When RINGO found the best matches, it searched for music you had not heard, then recommended it. Your musical pleasure was almost certain. And, if none of the matches were successful, saying so would perfect your string of bits. Next time would be even better.[13]

The founders of RINGO went commercial with the creation of Firefly, which launched its BigNote service in 1996. Users registered for a Firefly "passport" on the company's website and rank ordered artists on a list. The site then looked for those with similar tastes (which Firefly termed "trusted neighbors") and made recommendations. In addition, users could visit the home pages of their "trusted neighbors" on the Firefly website and engage in correspondence. One writer reported, "The concept was neatly logical: users would rate and review music, building a grand cross-referenced database of musical tastes. The more you told the system what you liked, the more Firefly would be able to make specific recommendations based on what *other* users liked."[14] However, Firefly was unable to develop a successful business model, and pieces of its technology were sold to Microsoft and Launch.com in 1997.[15] Microsoft wasn't especially interested in Firefly's collaborative filtering-recommendation engine; instead, it wanted Firefly's passport software for tracking user profiles, which became Microsoft's own Passport software for quick transfer of personal data.[16]

Amazon.com implemented the "Bookmatcher" feature on its website in 1997. Customers filled out forms indicating their interests, and the system matched them with book lists tailored to their entries. Bookmatcher required more time from users than most were willing to give, however, and Amazon now uses implicit data from previous purchases, which feed into its "New For You" and "Recommendations" features. Recommendations result from algorithms based on customer ratings and the buying patterns of customers who placed similar orders. To reduce errant data, users can exclude purchases from their "dossiers."[17] For example, a purchase of Boxcar Willie for your sister-in-law can be excluded from your goth- or techno-laden profile if you do not want the system to characterize you as an admirer of "the world's favorite hobo."[18]

An ambitious example of the human-based genre/mood matching systems, at least to judge from its name, is the "Music Genome Project" of a software company called SavageBeast (as in "music tames the . . ."). Its music experts sorted thousands of recordings according to a variety of attributes, or "genes," such as rhythm, lyrics, and instrumentation.[19] On October 31, 2001, SavageBeast introduced what it called the "Celestial Home Jukebox," a system for organizing files, automatically building playlists, and soliciting personalized recommendations. Among other things, the "organization" function scans one's collection, renames and retags files with metadata, and places them in playlists according to genre, instrumentation, mood, rhythm, tempo, vocal style, and so on. A user can also select "seed songs" from which the program will create playlists. The program also provides links to retailers.[20]

User interfaces vary among human-based services, but most make recommendations based on reactions to a series of song clips. Users of failed start-up Music Buddha created a "musical fingerprint" by choosing for each clip a recognizable genre such as jazz, rock, or classical, which they then narrowed into smooth jazz or heavy metal. Then they chose from a limited number of lifestyle or mood options offered on a menu, such as "tattoos and pool cues" or "celebration of women." Then they auditioned a selection of eight-to-ten-second song "hooks," as in radio call-out research (described below), and indicated whether these clips matched their preferences. At the end of this process, Music Buddha produced names of recordings it thought users might like and provided opportunities to buy them or add them to a "favorites" list.[21] The CEO of Music Buddha likened the result to "the mix tape that the boyfriend gives to the girlfriend. . . . That is the oh-wow moment we are trying to replicate."[22] Other systems, such as Media Unbound, iTunes, and MusicMatch, combine collaborative filtering with "expert" classification. MoodLogic.com allows users to "search by characteristics such as 'romantic R&B songs from the 1970s,' and visualize them graphically with elements called 'mood magnets'" that have been collaboratively identified by its users.[23]

Listening machines, on the other hand, are fully automated. They sort recordings according to their actual waveforms. In the CantaMetrix system, a computer analyzes a source track's waveforms for melody, tempo, harmony, timbre, and density and then retrieves similar recordings. Listeners may select a mood on the interface, such as "happiness," and stream preselected tracks. Such systems have long precedents. According to Lanza,[24] Muzak president Waddil Catchings originated the idea of assigning each song in the Muzak library a code that could be stored and transmitted according to rhythm, tempo, instrumentation, and ensemble size. The results were then piped into munitions factories during World War II to flatten out work-efficiency curves.

Problems with Personalization

To date, personalization systems have failed to realize the success predicted by their developers. A 2003 study by Forrester Research found that "of the 30 million people who used a personalized retail site [in 2002], only 22 percent found it valuable."[25] Such systems are plagued with inherent problems. One problem is the "cold start": when making matches, a company needs a large group of consumers who have made a large number of purchases before it can predict future choices. A related problem is the "popularity effect." As one executive said, "The holy grail [of CRM] is to be able to capture all the customer's interaction[s] in detail and get smarter about what *not* to recommend. . . . We can recommend very well. Knowing when not to bother someone is much harder."[26]

Thus, systems err on the side of false negatives (not offering music you might like) rather than false positives (offering music you might not like). This inherent conservatism results in predictable choices. According to one critic of the Firefly service,

> The service would rarely, if ever, break out of the mold of mainstream bands and recommend fringe music you'd never heard of before. And if your tastes strayed across numerous niches—say, you liked country and pop and techno, but weren't particularly devoted to any one genre, Firefly was equally problematic; the odds of finding a community of users with identically eclectic tastes were slim.[27]

A similar concern was raised about GroupLens, which has been used by Amazon.com, CDNow.com, and Music Boulevard. A music critic wrote,

> Just because I like a particular jazz musician such as Sonny Rollins—and I do—doesn't mean I want to hear everyone who sounds like him. Sometimes I want to hear music unlike anything I've heard before and no recommendation engine will be able to find those elusive musicians for me. I won't even be able to say what I want, or what I like, unless I hear it.[28]

These systems share the weaknesses of contemporary radio-market research practices such as call outs, wherein listeners evaluate five-second snippets, or hooks. Inevitably, given such a limited sample, listeners respond favorably to familiar music and unfavorably to unfamiliar or stylistically innovative music. Such systems make it unlikely that listeners will be exposed to unfamiliar genres and develop new tastes and interests. Filtering systems, similarly, can know only what the user already likes, and they are further restricted by the arbitrary parameters imposed by coders.

Human-based genre/mood classification, such as the Music Genome Project, is no less problematic. A comprehensive set of genre definitions and mappings is difficult or impossible to establish. Genre classification is an intensely subjective process, compounded by continuously proliferating and evolving categories that are not well-defined in the first place. And mood is even more subjective. For example,

> When MoodLogic's database was searched for all songs with the mood "aggressive," the system comes up with relatively mild fare. . . . CantaMetrix [in comparison] responded to the same word with snarling tracks by Black Sabbath and System of a Down. But when asked to find songs similar to Eric Clapton's recording of "Hoochie Coochie Man," CantaMetrix suggested a sweet song called "God Only Knows" by some female gospel singers called the Martins, and the country classic "I Saw the Light" by Hank Williams.[29]

Experts in music genres may be so narrowly knowledgeable that they sacrifice breadth to depth in their classification schemes. Untrained study participants can fail in detail and consistency. Yet many filtering systems for document-based sites—Google, Blogdex, Kuro5hin, Slashdot, and All Music Guide (AMG)—transform the judgments of such participants into aggregate recommendations.

The oldest, and probably most common, means of affecting such a transformation is the playlist, an early form of which can be found in the BBC program "Desert Island Discs," which dates from 1944. Playlists are popular among MSPs and clearinghouses. iTunes, for example, introduced a playlist function in October 2003, when it solicited playlists from musicians. iTunes now features "iMix," in which users publish playlists that can be rated by others, and "Party Shuffle," which automatically chooses songs from a user's library. Rhapsody's playlist function allows subscribers to e-mail their playlists to others; if the recipient also is a Rhapsody subscriber, he or she can click on an attachment that will play the sender's playlist. Rhapsody also offers customized playlists generated automatically when users enter three tracks in the hope of finding other, similar tracks. Yet user-provided playlists also present problems. Playlists from different services are incompatible, user behavior is inconsistent, and metadata may be missing or incomplete. While celebrity playlists or endorsements may serve a valuable function for marketers, they are constructed to support a public persona, and playlists may be abused by merchandisers of hard goods who want to get rid of excess inventory.[30]

Ultimately, filtering systems attempt the impossible. Tastes are not fixed—they are plastic and highly subjective—yet these systems are built on the assumption that tastes are objective, mechanical, and knowable, and that they are easily reduced to mathematical formulas. Such systems "[take] the stunning breadth of choices and [boil] them down to a limited number. In doing so, filtering fails to unearth the incredible diversity of our tastes, the quirkiness inherent to being human."[31]

In less artificial online communities—those created by members—recommendations may be shared more successfully. Through such means as "viral marketing," these communities may very well supersede music magazines, radio and television, and live performances as a means of publicizing artists and genres. This tendency already appears in "burning circles," in which members send thematic CD compilations to other members of the group.[32] Similarly, Napster enabled users to discover new content by browsing shared folders. But such coincidental collaborative filtering is unlikely with MSPs, clearinghouses, and even second- and third-generation network and service sites like KaZaA and Gnutella, which lack Napster's sense of community. As Napster morphed into Napster II, music fans decried the loss of community: "What I loved about Napster was the ability to connect, often in the wee hours of the morning, with total strangers who shared my tastes and interests and to discover new music, which I would never have heard otherwise."[33]

Critics contend that "iTunes is about music as a commodity; Napster was about music as mutual experience. iTunes is about cheap downloads; Napster was about file sharing—with sharing the key word."[34] In a cynical denouement, community filtering may come to resemble multilevel marketing schemes like Amway: each computer will become a vending machine, and each listener a potential sales representative for the industry, through incentives like free concert tickets to those who convince friends to buy a piece of music.

Despite its drawbacks, CRM is useful for cross-promoting the product lines of partner companies or subsidiaries. Mergers and acquisitions in the record industry make cross-promotion particularly lucrative, if personalization can identify customers who have affinities for products in both of the merged operations. The same software can adjust pricing and offer bargains. But CRM's greatest value may be its ability to identify the individual customer's value to the company or firm. Through CRM, one bank discovered that 20 percent of its customers created all of its revenues, while the other 80 percent were "destroying value."[35] The value of customer profiles traded between portals, affiliates, and advertisers is difficult to underestimate. Regulation places the only significant limits on data collection, since the Internet has few remaining technical limits for surveillance and profiling. Despite strong grassroots activism on privacy issues, the legal system has protected the interests of database vendors and financial companies. While consumers in Mexico, Canada, and the European Union enjoy personal-data protections, an electronic right to privacy has never existed in the United States for individuals. Consumers must assume that they have negligible rights to privacy in telephone or Internet communications.[36]

A single act of data mining for clues about our habits can give researchers only a one-dimensional profile of our identities. However, this profile can be enriched and blended with other sources of intelligence about us. As citizens turn over more and more privacy and independence to bureaucrats and marketers simply by making transactions as consumers of culture and information, the ubiquity of data mining threatens to transform U.S. society into a technocracy in which our role as citizens is completely subordinate to our roles as consumer clients for corporations and the state. These roles are defined by a "panoptic sort" made possible by increasing sources of information about our daily lives, as private or secret databases are sold, merged, and resold, and as information-retrieval technologies grow ever more sophisticated.[37] Each new data point gathered by the panopticon can enrich the complexity of the next sort and add to the strategic value of a database. As identity databases from private and public sectors merge, various one-dimensional definitions of us can be holographically expanded into multidimensional representations. But these representations, however thorough, remain hollow, reductive, and finally dehumanizing.

DIGITAL-RIGHTS MANAGEMENT

CRM is intended to build and track audiences, and DRM (which is central to the trusted-computing movement) is designed to regulate their online activities. Digital-rights management covers description (defining the content and its uses), identification (defining the user), and protection (ensuring "legitimate" use by "legitimate" users, as defined by the provider).[38] To accomplish these ends, DRM encodes licensing terms into digital files and tracks consumers' adherence to them during access and use. Minimal DRM authenticates users and provides secure (encrypted) file transmission, while robust DRM may include digitally watermarking copies, certifying content authenticity, and managing metadata attached to files. A comprehensive DRM system, such as Microsoft's Media Rights Manager, covers the entire lifecycle of products, from mastering to manufacturing to distribution to playback. As forensic evidence, DRM should leave a trail of checkoffs and documented assurances of identity and authenticity.

DRM Applications

To date, DRM technologies have followed three primary models. The first involves sealing content in electronic envelopes, which "legitimate" users can open by obtaining the appropriate codes, or "keys." These keys can be overt, in the form of passwords, or covert (embedded in the computer). This encryption process allows only authorized users to play back files, and it enables content to be priced according to "a fixed number of plays, unlimited plays for a limited time (with more time purchasable for additional cost) or selective access to different levels or dimensions of a product depending on the price paid."[39]

A second model involves developing algorithms for electronic watermarks or tags, which are directly inserted into files. Watermarks enable files to be traced back to their sources (files also can be identified by "hash marks" unique to each file, and "fingerprints," mathematical properties of a file's sound waves that can be measured and compared to a database).[40] When applied to music files, watermarks introduce subtle changes intended to thwart unauthorized duplication. They may repeat a particular frequency at specific intervals, remove specific tones in a narrow band, or add signals supposedly outside the range of hearing. Liquid Audio's DRM adds code during "dithering," or the insertion of low-level noise into digital recordings. The process is defined as follows:

> By "sculpting" the dither, Liquid Audio can . . . encrypt up to sixty-four characters, including the International Standard Recording Code (a

> sort of serial number for recorded sound), a second code identifying the computer that watermarked the song, a third identifying the computer that downloaded it, and a fourth, added at the time of the sale, giving information about who bought the song.[41]

Watermarking also allows providers to track the use of content. When a consumer downloads music from a website onto a CD recorder or portable player, paying for it with a credit card, the watermark can incorporate the user's name, credit card number, e-mail address, and other information, which might be stored at the record company's website.[42] The key benefit to the provider is "addressability," allowing the provider to sell the buyer upgrades and new products, or sue the buyer when he or she has engaged in unauthorized duplication. Early digital watermarking was performed by Philips CD recorders, which added a digital serial number to every CD they recorded so that CDs could be traced to the originating machine.[43]

A third, and increasingly common, type of DRM involves "locking up" data by eliminating record, cut, copy, paste, and print functions from applications, or limiting these applications to one computer.[44] Electronic texts may include copyright-protection systems that can "enforce rights such as 'print once only, store electronic copy for 90 days only, do not e-mail.'"[45] Microsoft's "Product Activation" technology also requires each user to register software over the Internet or by phone.[46] Digital identifiers are increasingly found in hardware as well as software: "[In 1999] Intel assigned a digital identifier, known as a processor serial number, to every new Pentium III chip, but disabled the feature a year later, after privacy groups said the serial number threatened to make anonymous Web surfing and Internet transactions impossible."[47] Hardware DRM examples include regional DVDs, inkjet printers built with chips that prevent refilled cartridges from being used, and cell phones with built-in manufacturer instructions to preclude the use of batteries from outside manufacturers[48] or switching cell carriers.

Although media companies "publicly admit that much of the piracy of major releases takes place before any consumers lay eyes or ears on the products,"[49] the incorporation of DRM systems into the Celestial Jukebox ensures that consumers receive only the products and services they pay for, and on the providers' terms. The first generation of DRM technologies sought merely to lock up content and limit use to purchasers, but more recent DRM involves the "description, identification, trading, protection, monitoring, and tracking of all forms of rights usages over both tangible and intangible assets including management of rights holders relationships."[50]

Different Standards

Rights-management technology precedes the advent of digital technology (Macrovision, an early leader in DRM, created a copy-protection system for VHS videocassettes). The industry expanded dramatically with the implementation of copy protection for computer diskettes and encryption systems for hardware. DRM for Internet distribution drew widespread attention from marketers and engineers soon after the Netscape Web browser was released; a major event was a January 1994 conference titled "Technological Strategies for Protecting Intellectual Property in the Networked Multimedia Environment."[51] Widespread file sharing and software copying in subsequent years increased the popularity of DRM. Intellectual-property-rights holders have pushed for incorporation of DRM "content protection" into DTV digital television signals, and the FCC dutifully imposed a DRM requirement for DTV tuners. However, the D.C. Circuit Court in the United States rejected the FCC's DRM requirement in 2005. As DRM-encoded CDs and DVDs flood the market, the technology is developing rapidly as a research area for hackers and attorneys.

From the outset, DRM for the music industry was intended to prevent CD tracks from being "ripped" by computers. The first CD issued in the United States with copy protection was the low-profile "A Tribute to Jim Reeves" by veteran country singer Charley Pride; it was released in May 2001 with no warning on its cover. It used technology from SunnComm, a Phoenix start-up, that directed a user, who placed the CD in a computer, to a website where he or she would register and download an encrypted replacement. SunnComm, however, rapidly faded amid allegations of stock manipulation and the departure of its chief technology officer.[52] It was succeeded by more successful ventures, such as Macrovision and Midbar. Macrovision's SafeAudio system took the draconian approach of corrupting the CD's actual audio signal, rather than merely its table of contents, as had been the case with earlier systems. The error-correction function of CD players allows the disc to be read, but computer drives, which lack such a function, either reject the disc altogether or produce mangled copies filled with pops and hiss. Error correction on CD players, however, was intended to address scratches, not intentional errors, so CDs encoded with SafeAudio are much less tolerant of scratches.

The Cactus Data Shield system from Midbar, an Israeli firm, also corrupts the audio signal, but in a different way:

> In this case blocks of audio are replaced with blocks of control data. A normal CD player ignores the control data and fabricates the sound of that block using its error recovery circuitry. . . . When the CD is copied using a computer or CD-to-CD copier, the control blocks are interpreted as audio, which means that the manufacturers can insert whatever sounds they wish into a copied recording, even sounds designed to damage speakers. The *New Scientist* ran an article on this copy-protection mechanism, based on research into the original patents, but it was forced to withdraw the article under pressure from Midbar.[53]

In June 2000, BMG issued a trial release of *Razorblade Romance* from a Finnish "love metal" band, HIM, that featured Midbar's system, but BMG was forced to recall the disc when a number of buyers couldn't play it.[54] Labels began releasing CDs with Macrovision DRM in early 2001, stealthily slipping them into stores. Citing nondisclosure agreements with record labels, Macrovision refused to say which CDs contained DRM; a spokesperson claimed, "They don't want to influence the listener's potential experience."[55] Macrovision held an edge in the DRM race because of the widespread application of its technology in DVDs. The company acquired Midbar in November 2002 and licensed its system to Microsoft in April 2003. By severely altering the open Red Book standard for CDs as they do, and in a proprietary way, these DRM schemes for music alter the music industry's main product line. Music labels may now sell these products as shiny discs with music on them—but not as real CDs.

SDMI and Its Aftermath

Content providers are caught in a dilemma in their attempt to supplant the Darknet's ubiquitous MP3s with a trusted format: incompatible DRM standards might hamper widespread adoption of digital technology, yet use of a single standard may result in a software monoculture controlled by one company. Moreover, the widespread adoption of a single standard undoubtedly would serve as a magnet for hackers, thereby rendering the standard useless. For example, the dominance of Microsoft's operating system has attracted swarms of hackers, who have plagued it with viruses and worms, while the Apple and Linux operating systems have remained relatively unscathed. DRM is particularly vulnerable to hacking, as the end-to-end operability of trusted systems is based on a long chain of control, which can be broken at any link.

Nevertheless, on December 15, 1998, the RIAA and the Big Five embarked on an effort to create a universal DRM system called the Secure Digital Music Initiative (SDMI). SDMI was a consortium of more than two hundred members with highly divergent and often antagonistic interests: record companies, consumer-electronics companies, and information technology companies. Its ranks included AOL, AT&T, IBM, Microsoft, Matsushita, Sony, RealNetworks, Liquid Audio, Samsung, ASCAP, Compaq, and Intel. Ironically, Napster was also a member, but no consumer or civil-rights groups were represented. SDMI's goal was to create a "gate" through which all digital content would pass, and which only the authorized user (the purchaser) would be allowed to open. It also would set an expiration date for use, limit the number of copies the purchaser could make, and trace those copies back to that purchaser. SDMI was a further development of the trend embodied in legislation like the 1998 Copyright Extension Act, but it provided a private solution to the public question of intellectual-property rights, and in the process it secures content protection into perpetuity, independent of any limits imposed by intellectual-property law.

The SDMI scheme encompassed all consumer-electronics and media-products producers. Here's how it was supposed to work. SDMI was to use two types of digital watermarks, which would be encoded in music in allegedly "nonaudible" ways. Electronics retailers would begin selling only SDMI-compliant hardware for computers, home stereos, and portables. Meanwhile, CD manufacturers would leave a "robust" watermark on all licensed or legitimate copies of a retail CD. The robust watermark would identify a CD as an "original" copy. The number of copies permitted by the copyright owners would be encoded in the watermark, and copying would be enabled for that specific number of iterations on an SDMI-enabled recorder. When copied by an SDMI-compliant device, the recorder would create a "fragile" watermark. CDs bearing fragile watermarks can play in SDMI-compliant devices (a portable player, car stereo, home stereo, etc.), but they cannot serve as an original source for new copies, as the SDMI device would refuse to record from them. After making a maximum number of permitted copies, consumers would have to repurchase the CD. In time, the SDMI standard could mutate in ways that would require music fans to purchase a separate copy of a piece of music for every player they owned.[56]

SDMI selected the watermarking system created by Verance Technologies as the standard (a system also selected by the 4C entity for DVD-Audio). On June 30, 1999, SDMI announced an agreement among its participants that would result in a two-step process. "Phase I" devices would play unprotected MP3 music as well as unprotected music based on SDMI technology. Once the recording industry had settled on a secure format for digital-music delivery, a PC with a Phase I software player would receive a signal that a software upgrade incorporating Phase II technology was available. SDMI noted that "the upgrade is not mandatory, but it is necessary in order to play new music that includes Phase II SDMI protection."[57] Phase II devices (scheduled for December 2000) would play MP3 music as well as SDMI-protected music, but it would screen incoming music and prevent devices from playing illegally obtained songs. In addition, Phase II software would restrict the number of copies a consumer could make from a protected CD or other source, to a number determined by the content's "usage rules."

An Exercise in "Vaporware"

Because of its ambitious agenda, SDMI quickly ran into development difficulties. Although watermarks were supposed to be undetectable, a panel of record producers in July 2000 found that the Verance watermarks were plainly audible on DVD-Audio CDs. One producer complained of a "high frequency buzzing," and noted that, "during complicated parts of the music, the sound became muddled."[58] After a disastrous London demo, an SDMI spokesperson admitted, "We are starting all over again."[59] The prospect of unrealizable "vaporware" hovered over the consortium's efforts. Thomson Multimedia SA (which owns RCA Electronics) announced that it intended to introduce products that lacked SDMI compatibility, although it retained membership in the organization. According to a Thomson spokesman, some of the record industry's expectations for SDMI were "off the wall."[60] Development lagged far behind schedule. In January 2001, the deadline for SDMI's implementation was pushed back to the following June, a date that would still allow manufacturers to sell compliant devices by Christmas 2001.

As development slowed, dissension swelled within SDMI's ranks. Micronas Semiconductors, a leading supplier of chips for MP3 players, abandoned SDMI in early 2001 because of the delays. The SDMI scheme was revised repeatedly; while the plan for robust watermarks was retained, the fragile watermark was abandoned out of concern that the growing popularity of broadband delivery would reduce the need for files to be compressed.[61] SDMI also was hobbled by the divergent goals of its participants. At the same time, some companies, including Sony, were already developing their own copy-protection plans. A number of differing DRM standards were already in place, causing difficulties with compatibility across platforms. For example, BMG had licensed competing technologies from Reciprocal and Digital World to process secure sales transactions. According to Gene Hoffman of eMusic, "It's akin to supporting three different word processor programs on a computer."[62] Dissension within SDMI's ranks led Leonard Chiariglione to resign as executive director in January 2001.

But perhaps the greatest threat to SDMI's intended hegemony was the fact that every protection scheme can be broken. As one industry observer noted, "The group believes it can do in less than a year what the entire computer software industry has been unable to do in two decades: stop software piracy."[63] Copy protection in computer software had largely failed. In the course of two years, for example, the leading producer of spreadsheet software, Lotus, had to cut prices and drop copy protection because it was losing sales to competitors.[64] An executive at Microsoft belittled the record industry's complaints: "The software industry loses more money to piracy than the record industry makes."[65] To hype its protection scheme, on September 15, 2000, SDMI offered a $10,000 award to anyone who could remove watermarks from a sample of recordings without degrading their audio quality. The "Hack SDMI" challenge provoked criticism from the hacker community, who saw it as "a corporate attempt to freeload off hacker brainpower" and who urged a boycott of the competition.[66] Yet, according to SDMI, 447 hackers took up the challenge. All four systems proposed for SDMI Phase II were successfully hacked by a team of researchers from Princeton University, Rice University, and Xerox PARC, headed by Edward Felten, a professor of computer science from Princeton. The team announced that SDMI's online "oracle" verification system had acknowledged their success and claimed that any reasonably sophisticated hacker could circumvent the technology.[67]

Every DRM system to date has been hacked, often by preexisting software. Both Macrovision's SafeAudio and Midbar's Cactus were defeated by software released in 1999—Clone CD, from the German firm Elaborate Bytes, which duplicates CDs bit by bit, rather than as files.[68] The very day that Windows Media Audio Version 4 was released in 1999, hackers created a workaround program called "unfuck.exe" that was promptly distributed on the Internet.[69] In May 2004, Jon Johansen cracked Apple's FairPlay DRM system. Johansen, who also was responsible for the DeCSS hack (described below), enabled iTunes downloads to be played on computers equipped with the Linux operating system.[70] A scandalous revelation of the emperor's nakedness occurred in October 2003, when a first-year Princeton graduate student, John Halderman, posted a paper analyzing SunnComm's MediaMax DRM system, which is used by BMG. Halderman described the basics of the system and explained that it could be disabled by simply holding down the Shift key when inserting a CD into a Windows PC.[71] He concluded that "anti-copy CD technologies will prove unfruitful, and will therefore eventually be abandoned by record companies."[72] SunnComm was not amused by the disclosure. After its stock value dropped $10 million (nearly a third of the company's worth), the company threatened to sue Halderman for reporting the discovery. In the wake of a public outcry, the company later backed down on the threat and expressed regrets for pursuing legal action.[73]

While the software, music, and film industries have embraced DRM, it remains highly problematic for hardware manufacturers, who seek user-friendly devices, and, of course, for consumers, who ultimately bear the costs of DRM research, development, and implementation. Julian Midgely of the UK Campaign for Digital Rights claimed, "The record companies are using customer's money to fund their own experiments, by throwing different systems onto the market to see who notices, who cares and what they can run with."[74] Other groups, including New Yorkers for Fair Use, New York Linux Scene (NYLXS), and the Free Software Foundation, lobby against

SDMI and other DRM initiatives on the rationale that "DRM is theft" of a public good by intellectual-property owners.[75]

Table 4.2 identifies some standards that are vying for dominance in online audio and video delivery.

DRM market leaders are Apple, Microsoft, IBM, and Intertrust.[76] These systems are plagued by incompatibility:

> Tracks downloaded from iTunes will only play on an iPod while tracks downloaded from rival services such as Napster, My Coke Music and MSN will not. And other portable music devices, such as those made by Creative, Rio and Sony, are not compatible with iTunes but do play tracks downloaded from dozens of rival services.[77]

Contributing to the compatibility problem, competing standards for "middleware" have slowed digital-music distribution across DRM platforms. XrML (eXtensible rights Markup Language), a software system that allows copyright holders to encode terms of use into a digital artifact, originated at Xerox PARC in the mid-1990s. Xerox received patents for XrML in 1996 and 1998 and transferred the patents to its ContentGuard spin-off in 2000. No doubt remembering that its concept of the graphical user interface later gave Apple its raison d'être, Xerox claimed, hyperbolically, that its patents gave it the right to "the very idea of a rights expression language."[78] ContentGuard licensed its technology to Sony in 2003 and was purchased outright from Xerox by Microsoft and Time Warner for $83 million in 2004.[79] The deal was under investigation in Europe in 2004 for possible antitrust violations. A primary competitor to XrML is XMCL (eXtensible Media Commerce Language), developed by

Table 4.2. Copy-Protection Systems

Developer	*System*	*Media*
Adobe	Content Server	Electronic Text
Apple	Fair Play	Audio
IBM	Electronic Media Management System	Audio
Microsoft	Windows Media Rights Manager	Audio, Video
RealNetworks	Helix DRM and Harmony	Audio, Video
Sony-Philips	Intertrust	Audio, Video, Software
Xerox	ContentGuard	Audio, Video

Source: John Markoff, "Five Giants in Technology Unite to Deter File Sharing," *New York Times*, 5 Jan. 2004, 1(C); and authors.

RealNetworks in 1999 as an open standard to define a common DRM framework in opposition to XrML. While both systems are intended to serve as umbrellas for DRM, RealNetworks asks for royalty fees from use, while Microsoft simply charges for software.[80] Another open-source product, Open Digital Rights Language (ODRL), is championed by Sun Microsystems;[81] yet another open-source product, Extensible Open RDF (EOR), does not use proprietary software or standards. RealNetworks' Harmony software initiative created a translator that allows most of these standards to work across multiple platforms, thereby enhancing their compatibility.

Microsoft, Real, and Apple: "Trusted Systems"?

Hardware-software synergies will strengthen as commercial digital distribution becomes mainstream. The computerization of consumer electronics has positioned information-technology companies to shape the new markets for music. Digital file-encoding schemes and digital-rights management standards (such as Intertrust) all exploit emerging hardware-software linkages. Sony's Connect service for ATRAC-encoded digital music files is downloadable to Sony portable music players; Apple's iTunes sends AAC-encoded (Advanced Audio Coding) songs to iPods, which currently drive demand for its online music business;[82] Microsoft's WMA system offers encrypted files for use on Windows-compliant computers and portables. While they do not own content, Apple's and Microsoft's diversification into DRM, music players, and online music stores indicates that these companies expect to play major roles in moving music from the wired Internet to portable players, cell phones, and video games while making money at every stage in the chain.

The licensing fees stemming from DRM patents promise to be substantial; as one observer noted, "At stake for the technology industry is a lucrative, ground-floor position in the next wave of e-commerce: the digital delivery of media products—music, e-books, and movies—via the Internet."[83] Microsoft enjoys a substantial advantage in DRM technology, since it provides the dominant operating system for PC users. Most major music labels are already releasing songs in the Windows Media format, which contains built-in DRM. Microsoft's long-term strategy has been to tie specific copies of content to specific PCs, which largely precludes the possibility of transferring content to other devices. The company embedded DRM in its Windows XP operating system, in which a Secure Audio Path program "authenticates" digital audio and video signals before they reach sound- and video-card outputs. If the signals don't pass, the content is scrambled:

> The system is designed to work behind-the-scenes, so that consumers aren't aware of any digital rights management. When the operating system accesses media files, noise is added so that if the audio is intercepted, it won't be usable. Once the file makes it through the hardware device and passes it to the Windows Media Player, the noise is removed and the file plays.[84]

Following its longstanding strategies for bundling software, and basking in its victory in the "browser war" with Netscape, Microsoft has included the Secure Audio Path with its operating system, freezing out other DRM suppliers. The arrangement allows unprecedented control for Microsoft: a ubiquitous platform, a secure system, and near-universal acceptance. And Microsoft extended this control through the recording industry in its Palladium project, also known as the "Next-Generation Secure Computing Base for Windows."[85] Palladium was widely envisioned to be packaged with Microsoft Longhorn, its next-generation Windows operating system. In October 2004, Microsoft sent a letter to record labels, urging adoption of a unified standard (devised by Microsoft, of course) that would attract consumers through additional multimedia content such as bonus tracks "as a quid pro quo for adding effective [DRM] into the consumer experience."[86]

To strengthen its position in DRM, Microsoft uses a dual strategy: securing patents that will be valuable in the long term, and gaining a foothold in new markets by giving away technology in the short term. To avoid wars over patents, Microsoft has scrambled to settle claims of copyright infringement. Intertrust filed suit against Microsoft in 2001 on grounds that Windows Media Player infringed on its copyrights for DRM and trusted-computing technology. The suit was settled in 2004, when Microsoft agreed to pay Intertrust $440 million.[87] Two weeks before the Intertrust settlement, Microsoft settled a similar suit with Sun Microsystems for $1.6 billion. In July 2004, Microsoft announced that "it planned to increase its storehouse of intellectual property by filing 50 percent more patent applications over the next year than in the previous 12 months."[88]

As it secures patents and buys off competitors, Microsoft intends to prime the pump for its systems by giving away licenses for their use. For example,

> In mid-January [2003], Microsoft unveiled a new toolkit that would let record labels create music CDs containing, along with the normal tracks, preripped Windows Media versions suitable for uploading to a buyer's MP-3 tape player or PC, but is protected by Microsoft's digital rights management (DRM) technology to prevent copying and swapping. The toolkit, the DRM license and the use of the Windows Media Audio format is free for the labels, despite Microsoft's $500 million investment developing what many analysts regard as the best DRM technology available today.[89]

Of course, this "free" offer would require labels to purchase support and upgrades. This process of acquiring patents and giving away licenses (at least initially) is intended to allow Microsoft to become the de facto standard, undercutting competitors and cornering the market; Palladium will enable it both to disable unauthorized content and to squeeze out competitive software. After Microsoft settled in 2001 with the United States in an antitrust investigation, RealNetworks filed an antitrust suit against Microsoft in December 2003, charging that Microsoft was trying to monopolize digital-media delivery by bundling Media Player with its Windows operating system. In March 2004, the European Commission levied a $603 million fine on Microsoft and required the company to offer a version of Windows stripped of Media Player and to open to competitors its code for creating software for business servers. In contrast, the 2001 settlement in the United States did not require Microsoft to unbundle software.[90]

Nevertheless, RealNetworks faces an uphill fight against Microsoft. To counter Microsoft's growing market share in DRM (particularly among Hollywood companies), RealNetworks attempted to strike an alliance with Apple in April 2004 by licensing Apple's FairPlay DRM system.[91] When Apple rebuffed these overtures, RealNetworks developed its Harmony system, allowing users to skirt Apple's iTunes service and place downloads from rival services on iPods. In this, however, RealNetworks may have run afoul of the DMCA, since it had to crack Apple's DRM system in order to ensure adaptability.[92] Apple's FairPlay DRM system was widely lauded as less onerous than its competitors upon its release. FairPlay allowed downloads to be burned into ten consecutive CDs; after that, their playback order must be rearranged. In addition, music could be copied onto three computers.[93] By May 1, 2004, Apple restricted copying to a maximum of seven CDs, and the system also detected and blocked similar playlists. A former iTunes user sued Apple in 2005, accusing FairPlay of violating his fair-use rights, and the iPod-iTunes DRM linkage as an "unlawful bundling and tying arrangement" that violates federal and California state laws by "suppressing competition, denying consumer choice, and forcing consumers to pay supra-competitive prices for their digital portable music players."[94] Despite the Internet's idealized promise of disintermediation, DRM adds another layer between vendor and purchaser, with an accompanying price hike for consumers.

The DMCA Clampdown

At the turn of the century, several court cases examined the legality of DRM. The first of these, the DeCSS/*2600* case, originated when Jon Johansen, a fifteen-year-old Norwegian, developed computer code in the fall of 1999 that would allow computers with the open-source Linux operating system to play DVDs. Such a program was unavailable on the market. Johansen cracked the DVD encryption system (CSS, or Content Scramble System) to create DeCSS, which was distributed over the Internet. Norwegian authorities seized Johansen's computer and charged him with copyright infringement,[95] but Johansen was acquitted of all charges in 2003. Courts in the United States were not so charitable toward such activities. In January 2000, eight Hollywood studios sued Eric Corley, the publisher of *2600* magazine, for publishing the DeCSS code on the magazine's website. In August 2000, federal judge Lewis Kaplan ruled against Corley and ordered him to remove the software code, as well as links to other sites containing the code, from the *2600* website.[96]

In addition to underscoring the illegality of distributing a computer program with potentially infringing use, Judge Kaplan's ruling made it illegal even to link to someone else's website that contained such material. In his ruling, Kaplan argued that "computer code is not purely expressive any more than the assassination of a political figure is purely a political statement."[97] Adrian Bacon, the publisher of *Linux News Online*, responded, "I think that Judge Kaplan does not know his head from his ass. Outlawing a site from linking to another site that has DeCSS is just plain wrong."[98] In protest, the DeCSS code was reprinted on T-shirts, schematics, bar codes, games, and videos. An especially imaginative effort involved reprinting the transcript of the trial riddled with errors; when assembled individually, the typos represented an encryption of the DeCSS code.[99] Kaplan's decision was upheld in November 2001 by a three-judge panel at the Second Circuit U.S. Court of Appeals, which ruled that the code was more "functional" than "expressive."[100] Corley and his attorneys decided not to appeal the decision.

Edward Felten, who headed the team that cracked SDMI's watermarks, had previously served as one of twelve government witnesses in the 1997 antitrust suit against Microsoft, where he testified that he had written a program that could easily separate Microsoft's browser from the operating system,[101] to discredit Microsoft's antitrust defense that the separation was impossible. When Felten announced that he intended to present the team's Hack SDMI findings at an academic conference in April 2001, he received a letter from SDMI urging him to "destroy" the paper and avoid discussing "confidential" information. If he refused, SDMI threatened to prosecute him for violating the DMCA.[102] Felten's team withdrew the paper from the conference but filed suit in a New Jersey federal court to allow it to be presented at future conferences and submitted for publication. The paper was presented at a computer security conference in August 2001,[103] and Felten's lawsuit was dismissed in federal court after the RIAA (who, with Verance, were the primary instigators within SDMI for action against Felten) dropped their threat. In a disingenuous statement, Cary Sherman of the RIAA stated, "We are happy that the court recognized what we have been saying all along: there is no dispute here. Professor Felten is free to publish his work."[104]

Other researchers fared less well under corporate pressure. Dimitri Sklyarov, a Russian cryptographer, was arrested and held without bail on July 16, 2001, the day after he presented a paper on decrypting Adobe's software for electronic books to a conference in Las Vegas.[105] The arrest stemmed from Adobe's complaint to the FBI that the program violated U.S. law. If convicted, Sklyarov would have faced a possible five-year jail sentence and $500,000 fine. Sklyarov was an employee of ElcomSoft, which sold the program (which was legal in Russia and most of the world) over the Internet. As Lawrence Lessig noted,

> Mr. Sklyarov himself did not violate any law, and his employer did not violate anyone's copyright. What his program did was to enable the user of an Adobe eBook Reader to disable restrictions that the publisher of a particular electronic book formatted for Adobe's reader might have imposed. . . . A blind person, for example, could use ElcomSoft's program to listen to a book.[106]

Sklyarov languished in the Las Vegas jail for eleven days. He was then moved, in handcuffs and shackles, to a federal prison in Oklahoma and was finally relocated to another prison in San Jose, California. After three weeks in custody, he was released on bail. A public-relations nightmare led Adobe to withdraw support for the case, but the Justice Department continued with prosecution. In December 2001, a federal court allowed Sklyarov to return to Russia and deferred charges for one year, and a federal jury acquitted Sklyarov a year later.[107] Sklyarov put the affair into perspective for *Newsweek*: "It's not that someone decided this Russian guy is bad and has to sit in jail. It's the money. In the U.S. everything is related to money."[108]

The DMCA does not bar the dissemination of information; rather, it specifically bars trafficking in "any technology, product, service, device, component, or part thereof, that . . . is primarily designed or produced for the purpose of circumventing a technological measure that effectively controls access to a (copyrighted) work."[109] The Felten case exemplifies the recording industry's reluctance and bad faith in negotiating with the public, and particularly with institutions of higher education, which could reveal hidden social costs of copyright reform and even alter the terms of the copyright grab in cyberspace. Companies have an obvious stake in promoting the efficacy of their copy-protection systems. Yet the threat of fines and jail time for challenging corporate claims of invulnerable software has had, and will have, a chilling effect on academic research. As Lessig notes, "Research into security and encryption depends upon the right to crack and report. Only if weaknesses can be discovered and described openly will they be fixed."[110] These and other lawsuits seek to undermine fair use, a legal doctrine that has also included the consumer's right to first sale. They hold that while you can loan or sell a book or CD to a friend, you cannot do the same with an electronic text; similarly, while you can make a backup copy of a CD for your own use, you cannot do so with a DVD. As the *New York Times* noted, "the [DMCA] now made it illegal not only to photocopy and sell a copyrighted book, but also to simply tell someone how to open and read that book without the publisher's authorization."[111]

SDMI was placed on indefinite hiatus in May 2001 when it announced that its partners couldn't agree on fundamental copy-control rules for nonportable devices such as PCs and networked stereo components.[112] SDMI died of a broken heart, split as it was between "hardware" (consumer electronics) and "software" (music and video) divisions, often within the same conglomerates, including Sony.[113] Nevertheless, DRM code is thriving, as it has been for years. The Audio Home Recording Act of 1992 required "Serial Management Copyright System" chips, which precluded second-generation digital copies, to be installed in digital-audio tape recorders. Such chips also are involved in the use of recent digital formats such as Super Audio CD (SACD) and DVD-Audio. Under the FCC's "Broadcast Flag" requirements, content providers were assured that every digital TV would carry circuitry to prevent recording of copy-protected programs. DRM also will be implemented as software code, based on private contractual agreements between copyright owners, hardware manufacturers, and citizens, with terms that could bypass any protections established by law. The industry's anticipated transition to streaming and cell phones and other low-cost, specialized "Internet appliances," based on closed rather than open architectures, will accelerate such implementation. These developments ensure that while perfect control is impossible, effective control is probable. As Lawrence Lessig states, "Locks can be picked, but that does not mean locks are useless."[114] DRM essentially holds that consumers have lost any claim to managing intellectual-property rights for themselves. DRM implicitly posits consumers as criminals, with "the presumption that we are guilty until proven innocent."[115]

IMPOSING SCARCITY ON KNOWLEDGE AND CULTURE

DRM underlies the entire value chain for the Celestial Jukebox. Though the failure of watermarking has slowed it, the lockdown remains in progress. If media industries are to sell products on the Internet, they must be able to restrict access to those products.[116] This lockdown can be accomplished through code, as in DRM, and supported by laws such as the Digital Millennium Copyright Act and the Sonny Bono Term Extension Act—laws created at the behest of copyright holders in closed debates in which no consumer or citizen interests were represented. Consumer rights are not considered in DRM technologies, and public policy here—as in so many areas of digital knowledge and culture—does little or nothing to prevent potential abuses by transnational corporations. With DRM, it has been noted, "If a right is not explicitly permitted, it has not been granted; that is, prohibited."[117] Librarians were among the first professional groups to fight for preserving traditions such as the First Sale Doctrine and fair use, which have been ignored by the DMCA and eviscerated from public life today. People are surprised to find out how much they have given up as a result of the DMCA and the deepening and extension of copyright. Contemporary laws and software code reflect an upending of decades of consumer protection, ranging from fair use to privacy. This pattern is consistent with the breakdown of the regulatory function of the state as a protector of consumer interest, described by Mosco,[118] and its replacement as an enforcer of intellectual-property rights.

Whatever their legal implications, DRM systems are disastrous for consumers. As one writer noted, "The only thing this watermarking system does is *not* let you play your music."[119] DRM hampers backup and restore operations; a user will have to purchase a separate copy of music for every player, and an entire music collection may be destroyed if a single player fails. Finally, locking up music so that only subscribers can hear it—a recurring dream of the Big Four—would actually prevent most listeners from hearing it in the first place and so wanting to purchase it. Jaron Lanier, a leading figure in the development of virtual environments, sums up the counterintuitive logic of DRM:

> In fact, the easier it is to copy music, the less of a threat piracy will become. When piracy gets easier, professional pirates will have less to offer. The only pirates left will be fans, and there are lots of ways to make money from fans. . . . The reason the Recording Industry Association of America and the labels are pushing anti-piracy laws and technologies has nothing to do with preventing piracy. They're doing it so that they can control the new digital music channels. To keep anyone else, like you, from sharing the power. They're doing it to rip you off. Period.[120]

Jones presents the case for the social benefits of "music that moves" with the consumer from home to car, work, gym, lunch, shopping, school, and home.[121] However, the "music that moves" from the Internet to a portable player is still constrained in many ways. Noninteroperability, closed standards, and infringements on consumers' fair use and privacy all defy consumers' needs. These obstacles to "music that moves" suggest that the culture industry, in its concern with successfully shifting from the sale of manufactured goods to the rental of intellectual property, and in its obsession with control over the gatekeeping, bundling, and distribution of these rights, cares exclusively for the requirements of business, and little or nothing for "consumer sovereignty" or public needs.

Web visionaries see the online universe governed one day by an "adaptive Web," which will conform to the needs of its users rather than vice versa.[122] The Celestial Jukebox has been constructed in profound indifference to those needs. It is the result of years of corporate research and development conducted in a deregulatory environment that favors the interests of corporations over citizens. The Celestial Jukebox strengthens norms of consumerism and quiescence; it has permitted the entertainment industry—the consciousness-making sector—to resolve its crises of production and consumption at our expense. CRM and DRM together subject us to unprecedented surveillance and manipulation. Intrusive online agents demanding personal information and pushing advertising and media on us have given us all the harried feeling of airport travelers, funneled through an enclosed media fun house where we are repeatedly required to identify ourselves, and our privacy and our property are exposed to search and seizure. To its greatest discredit, the Celestial Jukebox's legal infrastructure bypasses what feeble legal remedies remain as a corrective to a technocratic and overreaching corporate sector, by denying fair use for citizens, technical knowledge and free speech for researchers, and creative freedom for artists.

NOTES

1. Robert W. McChesney, *Corporate Media and the Threat to Democracy* (New York: Seven Stories Press, 1997), 30–31.
2. Steve Jones, "Music That Moves: Popular Music, Distribution and Network Technologies," *Cultural Studies* 16, no. 2 (2002): 213–32.

3. Jeremy Rifkin, "Where Napster Has Gone, Others Will Follow," *Los Angeles Times*, 21 Aug. 2000, reprinted at http://www.commondreams.org/views/082100-102.htm.

4. U.S. Department of State, *Copyright in an Electronic Age*, USInfo publication, n.d., http://usinfo.state.gov/products/pubs/intelprp/copyrt.htm.

5. Ronald Grover and Tom Lowry, "Online Music: Can't Get No . . . ," *Business Week*, 3 Sept. 2001, http://www.businessweek.com/magazine/content/01_36/b3747104.htm.

6. *USAToday.com*, "U.S. Probing Online Music Ventures," Reuters Limited, 7 Aug. 2001.

7. *USAToday.com*, "U.S. Probing."

8. Pamela McClintock, "Hatch Tells Labels to Speed Online Licensing," *Variety*, 12 Jan. 2001.

9. R. Hewitt Pate, "Statement by Assistant Attorney General R. Hewett Pate Regarding the Closing of the Digital Music Investigation," U.S. Department of Justice, 23 Dec. 2003, http://www.usdoj.gov/atr/public/press_releases/2003/201946.pdf.

10. Ina Fried, "Mac Sales Up 40 Percent," *CNet News*, 14 Apr. 2005, http://news.zdnet.co.uk/hardware/0,39020351,39195066,00.htm.

11. Oscar H. Gandy, *The Panoptic Sort: A Political Economy of Personal Information* (Boulder, CO: Westview Press, 1993).

12. John Reidl, Joseph Konstan, and Eric Voorman, *Word of Mouse: The Marketing Power of Collaborative Filtering* (New York: Warner Books, 2002).

13. Nicholas Negroponte and Pattie Maes, "Electronic Word of Mouth," *Wired*, Oct. 1996, http://www.wired.com/wired/archive/4.10/negroponte_pr.html.

14. Janelle Brown, "Personalize Me, Baby," *Salon.com*, 6 Apr. 2001, http://www.salon.com/tech/feature/2001/04/06/personalization/print.html.

15. Brown, "Personalize Me."

16. Lisa Guernsey, "Making Intelligence a Bit Less Artificial," *New York Times*, 1 May 2003, G5.

17. Susan Stellin, "E-Commerce Report," *New York Times*, 28 Aug. 2000, C8.

18. Boxcar Willie, "Boxcar Willie: The World's Favorite Hobo," 2004, http://www.boxcarwillie.com.

19. Don Clark, "New Web Sites Seek to Shape Public's Taste in Music," *Wall Street Journal*, 14 Nov. 2000, B1.

20. *Music Industry News Network*, "Savage Beast Technologies and Auditude Join Forces to Offer Complete Solution for Managing Digital Music," 1 Nov. 2001, http://mi2n.com/press.php3?press_nb=29246.

21. Brown, "Personalize Me."

22. Clark, "New Web Sites," B1.

23. Clark, "New Web Sites," B1; Fabio Vignoli, "Digital Music Interaction Concepts: A User Study" (paper presented at the Fifth International Conference on Music Information Retrieval, Barcelona, Spain, 10–14 Oct. 2004), http://ismir2004.ismir.net/proceedings/p075-page-415-paper152.pdf.

24. Joseph Lanza, *Elevator Music: A Surreal History of Muzak, Easy-Listening, and Other Moodsong* (New York: St. Martin's Press, 1994).

25. Zachery Kouwe, "Getting Book Suggestions Online," *Wall Street Journal*, 29 July 2003, D2.

26. Guernsey, "Making Intelligence," G5.

27. Brown, "Personalize Me."

28. Frank Jossie, "Net Perceptions Hits the Big Time," Interactive Pioneers, n.d., http://www.interactive-pioneers.org/net_perceptions.html.

29. Clark, "New Web Sites."

30. Kouwe, "Getting Books," D2.

31. Larry Smith, "We Heard It through the Napster Grapevine—All of It," *Los Angeles Times*, 10 Sept. 2000, M5.

32. David Gallagher, "For the Mix Tape: A Digital Upgrade and Notoriety," *New York Times*, 20 Jan. 2003, G1.

33. Henry Jenkins, "Playing Our Song? Digital Renaissance," *Technology Review*, 2 July 2003, http://www.mittechnologyreview.com/articles/03/07/wo_jenkins070203.asp.

34. Shawn Yeager, comment posted to Pho listserve, 7 July 2004, http://www.pholist.org.

35. Simon London, "How to Know What the Customer Wants Next," *Financial Times*, 13 July 2001, 10.

36. Jean L. Camp, *Trust and Risk in Internet Commerce* (Cambridge, MA: MIT Press, 2001).

37. Gandy, *The Panoptic Sort*.

38. Jon Webb, "Digital Rights Management or DRM—Does (It) Really Matter?" white paper of the International Association of Scientific, Technical and Medical Publishers, 2003.

39. Tony Feldman, *Introduction to Digital Media* (New York: Routledge, 1997), 65.

40. Charles Mann, "Napster-Proof CDs: The Music Industry's Secret Plan to Safeguard Popular Music from the Wild Web," *Inside.com*, 27 Mar. 2001, reprinted at http://www.salon.com/tech/inside/2001/03/27/cd_protection/index.html.

41. Charles Mann, "Who Will Own Your Next Idea?" *Atlantic Monthly*, Sept. 1998.

42. Jon Pareles, "Trying to Get in Tune with the Digital Age," *New York Times*, 1 Feb. 1999, C6.

43. Sara Robinson, "Recording Industry Escalates Crackdown on Digital Piracy," *New York Times*, 4 Oct. 1999, C5.

44. Marion Long, "Fighting the Good Fight: Enforcing Copyright Law on the Web," *Inter@ctive Week*, 10 Sept. 2000, reprinted at http://smccd.net/accounts/karas/fighting.html.

45. Webb, "Digital Rights Management."

46. Robert Lemos, "Consumer Rights on the Block?" *ZDNet*, 17 Jan. 2001, http://www.zdnet.com/zdnn/stories/news/0,4586,2675033,00.html?chkpt=zdnn01171.

47. Chris Gaither, "Intel Chip to Include Antipiracy Features," *Boston Globe*, 10 Sept. 2002, C3, reprinted at http://www.xenoclast.org/free-sklyarov-uk/2002-September/003250.html.

48. Hal Varian, "Economic Scene," *New York Times*, 4 July 2002, C2.

49. Bill Rosenblatt, "2003 in Review: DRM Technology," *DRMWatch.com*, 31 Dec. 2001, www.drmwatch.com/drmtech/article.php/3294391.

50. Renato Iannella, "Digital Rights Management (DRM) Architectures," *D-Lib Magazine*, June 2001, http://www.dlib.org/dlib/june01/iannella/06iannella.html.

51. Bill Rosenblatt and Gail Dykstra, "Integrating Content Management with Digital Rights Management: Imperatives and Opportunities for Digital Content Lifecycles," Giantsteps Media Technology Strategies, 14 May 2003, http://www.drmwatch.com/resources/whitepapers/article.php/3112011, 4.

52. John Borland, "Merger Could Spawn More Copy-Proof CDs," *CNet News*, 5 Nov. 2002, http://news.com.com/2100-1023-964590.html.

53. Jim Peters, "Buying a New CD? Watch out for Inferior Imitations," Oct. 2001, http://ukcdr.org/issues/cd/overview.shtml.

54. Mann, "Napster-Proof CDs."

55. John Borland, "Copy-Protected CDs Quietly Slip into Stores," *CNet News*, 18 July 2001, http://news.com.com/2100-1023-270164.html?legacy=cnet.

56. Matt Richtel, "Standards Are Set for Thwarting Music Pirates," *New York Times*, 14 July 1999, C5.

57. Secure Digital Music Initiative, "Frequently Asked Questions," http://www.sdmi.org/FAQ.htm.

58. Thomason, Robert, "Record Labels Try to Remix the CD," *Washington Post*, 18 Aug. 2000, E1.

59. *New Scientist*, "Starting Over," 22 July 2000, http://www.newscientist.com/article/mg16722483.300.html.

60. Junko Yoshida and Margaret Quan, "MP3 Chip Maker Micronas Bails Out of SDMI," *EETimes.com*, 5 Jan. 2001, http://www.eetimes.com/story/OEG20010105S0006.

61. Roger Parloff, "Gung Ho for Christmas 2001, Secure-Music Initiative Streamlines Its Mission," *Inside.com*, 26 Jan. 2001, reprinted at http://www.bluespike.com/news-inside012601.html.

62. Matt Richtel, "New Economy," *The New York Times*, 17 Apr. 2000, C4.

63. Peter Lewis, "Internet Music, To Go," *New York Times*, 24 Dec. 1998, D3.
64. Hal Varian, "Economic Scene," *Wall Street Journal*, 27 Jan. 2000, C2.
65. Eben Shapiro, "Race Is On to Foil E-music Pirates," *Wall Street Journal*, 22 Jan. 1999, B4.
66. Janelle Brown, "Is the SDMI Boycott Backfiring?" *Salon.com*, 3 Oct. 2000, http://www.salon.com/tech/feature/2000/10/03/hacksdmi_fallout/.
67. Edward Felten, "Statement Regarding SDMI Challenge," http://www.cs.princeton.edu/sip/sdmi/announcement.html.
68. Tony Smith, "Old Code Defeats New CD Anti-Ripping Technologies," *The Register*, 8 Oct. 2001, http://www.theregister.co.uk/content/54/20947.html.
69. John Borland, "Hacker Cracks Microsoft Anti-Piracy Software," *CNet News*, 19 Oct. 2001, http://news.cnet.com/news/0-1005-200-7590303.html?tag=prntfr.
70. Andrew Orlowski, "iTunes DRM Cracked Wide Open for GNU/Linux. Seriously," *The Register*, 1 May 2004, http://www.theregister.co.uk/content/6/34712.html.
71. John Borland, "Shift Key Breaks CD Copy Locks," *CNet News*, 7 Oct. 2003, http://news.com.com/2102-1025_3-5087875.html.
72. John Halderman, "Analysis of the MediaMax CD3 Copy-Prevention System," 6 Oct. 2003, http://www.cs.princeton.edu/~jhalderm/cd3/.
73. Josh Brodie, "Threat of Lawsuit Passes for Student," *Daily Princetonian*, 10 Oct. 2003, http://www.dailyprincetonian.com/archives/2003/10/10/news/8797.shtml.
74. Barry Fox, "NSync CD Is Copy Protection 'Experiment,'" *New Scientist.com*, 2 Oct. 2001, http://www.newscientist.com/news/news.jsp?id=ns99991367.
75. Ruben Safir, "DRM is Theft: New Yorkers for Fair Use Go to Washington," *The Linux Journal*, 29 July 2002, http://www.linuxjournal.com/article/6243.
76. Figueiredo, Daniel, "Digital Rights Management Comparison and Selection Guide," Faulkner Information Services, 2003, reprinted at http://www85.homepage.villanova.edu/timothy.ay/DIT2165/00018503.pdf.
77. Owen Gibson, "Music Industry Presses for Common Standard for Downloads," *The Guardian*, 8 Oct. 2004, http://www.guardian.co.uk/arts/netmusic/story/0,13368,1322797,00.html.
78. Neil McAllister, "Freedom of Expression: Emerging Standards in Rights Management," *New Architect*, Mar. 2002, http://www.newarchitectmag.com/documents/s=2453/new1011651985727/index.html.
79. Bill Rosenblatt, "Time Warner Takes Stake in ContentGuard," *DRMWatch.com*, 8 Apr. 2004, http://www.drmwatch.com/drmtech/article.php/3337691.

80. Amy Harmon, "RealNetworks Goes after Bigger Piece of Media Library Pie," *New York Times*, 20 June 2001, C2.

81. Russell Kay, "XRML, the Language of DRM," *Computerworld*, 23 June 2003, http://www.computerworld.com/governmenttopics/government/legalissues/story/0,10801,82382,00.html.

82. Apple, "iTunes Music Store Begins Countdown to 100 Million Songs," 1 July 2004, http://www.apple.com/pr/library/2004/jul/01itunes.html.

83. Shapiro, "Race Is On."

84. Brad King, "Windows XP Can Secure Music," *Wired.com*, 13 Feb. 2001, http://www.wired.com/news/technology/0,1282,41614,00.html.

85. Amy Carroll et al., "Microsoft 'Palladium': A Business Overview," Microsoft Corporation, Aug. 2002, unpublished paper.

86. Tony Smith, "Microsoft Tells Music Biz to Back Lock-Down CD Standard," *The Register*, 16 Sept. 2004, http://www.theregister.co.uk/2004/09/16/ms_cd_copy_protection.

87. Steve Lohr, "Microsoft Settles InterTrust Suit for $440 Million," *New York Times*, 12 Apr. 2004, C1.

88. Steve Lohr, "Pursuing Growth, Microsoft Steps up Patent Chase," *New York Times*, 30 July 2004, C3.

89. Joe Wilcox, "Microsoft Protecting Rights—or Windows?" *CNet News*, 2 Feb. 2003, http://news.com.com/2100-1023-983017.html.

90. Steve Lohr, "Paring Away at Microsoft," *New York Times*, 25 Mar. 2004, reprinted at http://classwork.busadm.mu.edu/Economics%20Newspaper%20Articles/Microeconomics/Taxes,%20Regulations%20and%20Market%20Controls/2004/2004%2003%2025%20Paring%20away%20at%20Microsoft.PDF.

91. John Markoff and Steve Lohr, "RealNetworks Seeks Musical Alliance with Apple," *New York Times*, 15 Apr. 2004, C1.

92. Nick Wingfield and Pui-Wing Tam, "RealNetworks Seeks to Unlock iPod from iTunes," *Wall Street Journal*, 26 July 2004, B1.

93. David Pogue, "Apple's New Online Music Service," *New York Times*, 1 May 2003, G1.

94. Peter Cohen, "iTunes User Sues Apple over FairPlay," *PCWorld.com*, 7 Jan. 2005, http://www.pcworld.com/news/article/0,aid,119213,00.asp.

95. Amy Harmon, "Free Speech Rights for Computer Code?" *New York Times*, 31 July 2000, C1.

96. Lee Gomes, "Ruling in Copyright Case Favors Film Industry," *Wall Street Journal*, 18 Aug. 2000, B6.

97. Steven Bonisteel, "Appeals Court Gets Fodder for DeCSS Free Speech Debate," *Newsbytes*, 31 May 2001, reprinted at http://www.findarticles.com/p/articles/mi_m0NEW/is_2001_May_31/ai_75198112.

98. Declan McCullagh, "Only News That's Fit to Link," *Wired.com*, 23 Aug. 2000, http://www.wired.com/news/politics/0,1283,38360,00.html.

99. David Hamilton, "Banned Code Lives in Poetry and Song," *Wall Street Journal*, 12 Apr. 2001, B1.

100. Steven Bonisteel, "2600 Magazine Seeks Another Opinion in N.Y. DeCSS Case," *Newsbytes*, 14 Jan. 2002, reprinted at http://www.parentsurf.com/p/articles/mi_m0NEW/is_2002_Jan_14/ai_81788530.

101. David Hamilton, "Professor Savors Being in Thick of Internet Row," *Wall Street Journal*, 14 June 2001, B1.

102. David Hamilton, "Music-Industry Group Moves to Quash Professor's Study of Antipiracy Methods," *Wall Street Journal*, 24 Apr. 2001, A2.

103. Mike Musgrove, "Digital-Music Code Crackers Tell All," *Washington Post*, 16 Aug. 2001, E3.

104. John Schwartz, "Two Copyright Cases Decided in Favor of Entertainment Industry," *New York Times*, 29 Nov. 2001, C4.

105. Jennifer Lee, "U.S. Arrests Russian Cryptographer as Copyright Violator," *New York Times*, 18 July 2001, C8.

106. Lawrence Lessig, "Jail Time in the Digital Age," *New York Times*, 30 July 2001, A21.

107. Matt Richtel, "Russian Company Cleared of Illegal Software Sales," *New York Times*, 18 Dec. 2002, C4.

108. Stephen Levy, "Busted by the Copyright Cops," *Newsweek*, 20 Aug. 2001, 54.

109. Roger Parloff, "By Picking on Academics, Record Industry Plays the Bad Guy—at the Worst Possible Time," *Inside*, 25 Apr. 2001 (parentheses in original).

110. Lessig, "Jail Time."

111. Harmon, "Free Speech Rights."

112. Mark Lewis, "SDMI Pushed into Dormancy by Disagreements, Not Watermark Vulnerabilities," *Webnoize*, 29 May 2001.

113. Chris Partridge, "Copyright Wars," *vnunet.com*, 17 Mar. 2003, http://www2.vnunet.com/Features/1139612.

114. Lawrence Lessig, *Code and Other Laws of Cyberspace* (New York: Basic Books, 1999), 57.

115. Wendy Grossman, "To Protect and Self-Serve," *Scientific American*, Mar. 2001, p. 31, http://sciam.com/article.cfm?chanID=sa006&articleID=000B17E8-7A09-1C70-84A9809EC588EF21.

116. Camp, *Trust and Risk*.

117. Gartner G2 and the Berkman Center for Internet & Society, "Copyright and Digital Media in a Post-Napster World," Harvard Law School, 2003, http://cyber.law.harvard.edu/home/2003-05, 37.

118. Vincent Mosco, *The Pay-per Society: Computers and Communication in the Information Age: Essays in Critical Theory and Public Policy* (Toronto: Garamond Press, 1989).

119. Brown, "SDMI Boycott."

120. Jaron Lanier, "Making an Ally of Piracy," *New York Times*, 9 May 1999, 50.

121. Jones, "Music That Moves."

122 Thomas Gray, quoted in Barry Wellman, "Designing the Internet for a Networked Society," *Communications of the ACM* 45, no. 5 (May 2002): 96.

5

Digital Capitalism, Culture, and the Public Interest

The recording industry's lawsuits against some of its best customers illustrate the confusion and suspicion surrounding its most serious crisis in decades. The Napster decision emboldened the industry to pursue specific legal and technological solutions to the loss of revenues in retail outlets and the challenge of the Darknet. It has shifted public discourse about the Internet from the rights of consumers to the rights of the music industry, and it has encoded its template into every media e-commerce experience. The Big Four's adoption of technocratic software systems for online distribution, their divestiture from the prototype music-service providers, and their subsequent "outsourcing" of catalogs to clearinghouses were all strategies for reducing their actual risk in developing the Jukebox's infrastructure while at the same time establishing new markets for their intellectual property in cyberspace. Freed of the need to maintain delivery systems, the Big Four are content to reap the profits from their catalogs. The music industry has fallen back on an old strategy, well rehearsed in its industrial history, of resistance to disruptive technologies, which gives way to experimentation and dalliances, and then finally to their subversion or takeover.

The industrial dynamics that produced the Celestial Jukebox have determined its technology: lockdown through DRM and surveillance through CRM. The music industry hopes to choke off the radical potential of open networks and replace them with closed, proprietary systems. It hopes to arrive at a single, or at least a dominant, standard for both DRM and CRM. CRM is available commercially on various platforms, and many companies "roll their own" or augment outsourced applications with their own personalization software. CRM is growing more robust and better at surveillance, collaborative filtering, and reporting ("business analytics") for different networks. When CRM techniques become standardized, they will likely develop variations of Nielsen and Roper reports, with detailed real-time information on consumer behavior. If all goes as planned, CRM and DRM will reduce the user's music collection into a database, or favorites list, owned by the provider. The list is not really a list of recordings in a collection but an enumeration of temporary and restricted rights to access files in an online database operated by an e-commerce company.

Fans of music will suffer. They lose rights to first sale and, forced to repurchase access rights at intervals, become tenants in the hands of rent-seeking cultural landlords. In the absence of tangible commodities such as sheet music or CDs, the support structure itself (in this case, cyberspace) becomes the commodity; it will be controlled not by the user but by the recording industry, creating value through transactions or the process of circulation. One observer describes music on the Internet as returning to an intangible essence in which it "would stop being something to collect and revert to its age-old transience: something that transforms a moment and then disappears like a troubadour leaving town."[1] But this virtual troubadour will come or go, sing or not, at the behest of the Big Four.

THE PENDING TAKEOFF OF THE CELESTIAL JUKEBOX

The culture industry is poised to impose the Celestial Jukebox on all manifestations of mass culture. Virtual libraries at public colleges and universities have adopted DRM to lock down the contents of online journals, e-books, and other digital databases served by content-management software. Already, software, news, and other publications are available online by subscription or pay-per download. Wireless media such as cell phones are increasingly able to serve as a means to distribute digital content: ring tones became popular downloads before MP3s followed. Digital television systems carry high-quality audio signals and, when combined with interactive data broadcasting, can provide Jukebox-style video clips via "cellevision." Video distribution through the Jukebox model has begun in earnest on Web stores for video downloads, such as Movielink, a partnership between MGM-Sony, Paramount Pictures (Viacom), Universal Studios, and Warner Brothers Studios. Movielink downloaders purchase a twenty-four-hour virtual "ticket" to view a downloaded movie file. Limitations on infrastructure still prevent fast and reliable streaming of movies in high quality, but these limitations will be overcome with advances in signal compression and the growing ubiquity of wired and wireless broadband networks.

In the Celestial Jukebox, the old market-based economy of buyers and sellers is largely replaced by a new network-based economy of clients and servers. Rifkin claims, "In markets, the parties exchange property. In networks, the parties share access to services and experiences. . . . [They will form] a new kind of economic system based on network relationships, 24/7 contractual arrangements and access rights."[2] In such a system, value is not an inherent character of the product; it is the manner in which it reaches the consumer, and that, unlike a tangible product, can be easily sold to the same consumer again and again. The commodification of the popular-music product depends in large part on building and maintaining a one-to-one relationship between the music fan and the music service. In seeking complete control over this system, the culture industry seeks not to protect its properties from theft but to convert them from sales to rental and thereby transform us from owners to tenants.

Changes in the Music Industry Value Chain

The economics of online music distribution still depends in important ways on the institutions of radio broadcasting for marketing and promotion. As Frith notes, between 1920 and 1950, "The radio and music industries developed a symbiotic relationship. Music on record became the basis of radio programming; radio play became the basis of record selling."[3] Between 1992 and 2001, releases by the major labels dominated the charts for the leading radio formats, including contemporary hit radio (95 percent), rock (85 percent), and country (90 percent).[4] Radio-industry consolidation ran parallel with music-industry consolidation. Since the Big Five dominated radio airplay, "signing to a major label has become an almost necessary step to getting one's song played on the radio."[5]

Like radio, MSPs and clearinghouses hold the potential to meet consumers' desire to access any music, anywhere, at any time. Unlike radio, they have been plagued by incompatibilities among encoding schemes and player types. And yet, despite these problems, commercial digital distribution of music has begun to take off, supported legally by the statute and case law discussed in chapter 4. The model is preparing to adapt to wireless portables. Wireless telephony already accommodates MP3 downloads and polyphonic ring tones. As wireless broadband achieves broader coverage, WiFi and third-generation-enabled ("3G") cell phones and PDAs can incorporate a broadband data channel for Internet access, allowing these devices to receive streams of digital audio and video. These capabilities will allow MSPs and clearinghouses to compete, upon its diffusion in North America, with terrestrial digital-audio broadcasting (DAB) and satellite-radio services, a market presently shared by XM and Sirius. These events move us even closer still to the Celestial Jukebox model of pay-per or subscription service, tiering, and price discrimination.

In some ways, to be sure, the economics of online music distribution differs significantly from that of radio. The record industry acts as a gatekeeper for commercial music-radio programming, but radio merely provides promotion, whereas downloading and streaming provide distribution. None of the majors owns U.S. radio stations;[6] radio makes its money from advertisements rather than direct sales and provides only an ephemeral copy of music. While online services and digital broadcasters may exclude nonpaying audiences and even use price discrimination by charging certain customers more than others, terrestrial analog-radio broadcasters cannot. These defining characteristics give the subscription and pay-per-use models considerable advantages over the broadcasting model. A Celestial Jukebox operator need not fill its telecommunications channels in order to keep all its content online and always available. Live, on-demand transactions—whether sent by the consumer or "pushed" by the Jukebox—offer more profitability through premium pricing and shaping of demand.

Internet distribution might, in theory, lower artists' barriers to entry and increase audience's access to content as production and distribution become more accessible and inexpensive. It might, in theory, reduce the dependency on the major labels for financing, manufacturing, warehousing, shipping, and marketing that has kept so many artists both poor and frustrated for decades. The Pew Internet & American Life project surveyed musical artists about the Internet as a distribution platform and found enthusiasm for its opportunities and a willingness to look beyond the major labels' overriding concerns about file sharing:

> The first large-scale surveys of the Internet's impact on artists and musicians reveal that they are embracing the Web as a tool to improve how they make, market, and sell their creative works. They eagerly welcome new opportunities that are provided by digital technology and the internet. At the same time, they believe that unauthorized online file sharing is wrong and that current copyright laws are appropriate, though there are some major divisions among them about what constitutes appropriate copying and sharing of digital files. Their overall judgment is that unauthorized online file sharing does not pose a major threat to creative industries: Two-thirds of artists say peer-to-peer file sharing poses a minor threat or no threat at all to them.[7]

The Pew study does not distinguish between the attitudes of independent musicians and those affiliated with major labels. Independent artists stand to benefit from reduced recording costs on home-studio systems; they may keep more money from company advances or forego them altogether and begin earning royalties from sales sooner after the release date. By minimizing the role of record companies, artists can come closer to realizing their freedom of expression, no longer tied to contractual obligations with labels that ration material in marketable doses to maximize demand. Online digital delivery allows for infinitely varied packaging and releasing. A release can contain one or one hundred songs, and any schedule for release is possible. A song can be released the day it is recorded, and collections can be released in standard and advanced packages, akin to "director's cuts."

Direct availability of their work to consumers also frees artists from having to subsidize marketing; an artist who sells 70,000 records independently makes more money than one who sells 300,000 on a major label. In an early experiment, Prince released the triple-CD retrospective *Crystal Ball* primarily over the Internet. It sold 250,000 copies and earned $5 million in revenues, of which Prince kept 95 percent.[8] Ween issued two CDs of live recordings on their website after leaving Elektra Records. One of the members told the *New York Times*, "On Elektra our best-selling record sold 200,000 copies or more. But we still owe all this money to the label. Then we sold just five or ten thousand CDs through our Web site and raised $100,000 to make our new album."[9] Early experiments in bypassing the major labels show the potential for a very different sort of Jukebox. Several performers have employed the Internet to develop a patronage model based on subscriptions and serialized content. One of the earliest of these experiments, Todd Rundgren's "PatroNet," charged fans a sliding scale for packages of downloadable music, videos, and chapters of a book in progress.[10] Prince, Rundgren, and the other pioneers demonstrated that performers can base their own revenue-generating activities on fan-club operations, featuring presale concert tickets, meet-and-greet sessions for fans, and additional promotion through text messaging and e-mail.

As these pioneers became independent of their former labels, they were free to experiment with Internet delivery. Meanwhile, artists who are signed to major labels, and whose songs are migrating to online music stores, are beginning to demand contract reforms that address the lower costs of providing music online. Artists' royalties are typically computed from a base price per CD, which is calculated by making deductions from the suggested retail list price. These deductions frequently still include packaging, breakage, and "new technologies"—concerns about risk left over from the days of LP recordings. The Future of Music Coalition argues that record companies unjustly benefit from these adjustments, and it urges that "major label artists should demand to be compensated for these sales at a *unique* and *higher* rate."[11]

To date, however, no artist has achieved widespread popularity solely on the Internet. In fact, with lower barriers to entry, and cultural supply exceeding demand by an ever-greater margin, advertising and publicity will become, if anything, more important. New songs and artists will need, more than ever, the traditional marketing that has always been one of the Big Four's great strengths. As the recording industry focuses increasingly on the Internet, independent artists are likely to appear only at the extremes of the charts: either as "superstars" who, having been popularized by major record companies, are then able to draw revenues through independent fan-club operations; or experimental, innovative artists, who essentially give their work away to fans or bundle it through "branded" labels specializing in small, limited-appeal genres, such as Blue Note Records.

The rest, however, will rely on the promotion of the majors; and the hard goods that were once the object of such promotion are likely to become, instead, merely part of the means. At the moment, to be sure, it works the other way around: digital files now serve, like radio, as advertisement for tangible goods, which new technologies, such as DVD, make more desirable and expensive. Or they serve as a means of acquiring virtual goods (for example, Metallica offered bonus live MP3s to those who purchased their *St. Anger* CD and obtained an access code).[12] In the future, however, e-commerce merchants may adopt big-box loss-leader strategies to their online stores (like Amazon), essentially giving away files in order to attract customers to hard goods like clothing and soft drinks, or to other service packages. Apple has used this strategy with its iTunes music service, successfully bundling iTunes with Pepsi Cola. The long-term value of music copyrights may be in enabling commerce, not in serving themselves as end commodities.

To compensate for this lack of hard goods, and to rationalize their disavowal of fair use and intrusion into privacy, digital music providers tout greater selectivity, personalization, and community as "value-added" features to consumers, ever more necessary as consumers pick and choose single releases, rather than bundled collections, from a growing volume of material. These features are both collective (viral marketing) and individualized (CRM). Viral marketing is already found in the playlist sharing of Rhapsody and Napster and alternative Jukeboxes such as Weedshare, which has experimented with viral marketing as a revenue model for pay P2P. The marketing implications are obvious: "Through viral marketing, each PC becomes a vending machine, and each listener a potential sales representative for the industry, which would offer incentives such as concert tickets or credits for future purchases to those who convince friends to buy a piece of music."[13]

The personalization of CRM and the exclusiveness of DRM further media trends, away from broadcasting and toward narrowcasting, that have long driven cable and satellite television and that now drive satellite radio. You have to pay for access, and you get something ever narrower; eventually, with the Celestial Jukebox, you get a "channel" heard only by yourself. Such microsegmentation, however, has social costs. As one observer noted, "The music may be speaking right to me, but it's alienating being a niche market of one."[14] The Celestial Jukebox creates "audiences" by isolating its users and reaggregating them into a manufactured community of atomized streamers and downloaders.

The purpose of this political economy of the Celestial Jukebox is to encourage more consumption, faster production cycles by artists, and more disposable music. You use it once and throw it away; you are thinking about, and paying for, not this or that song or artist, but the virtual machine that delivers a reliable stream of musical pleasure to you alone. There is a down side that comes with the operation of this new system. Aside from the corporate-controlled "communities" enabled by CRM, the Celestial Jukebox lacks common spaces and public forums for sharing tastes and experiences, further alienating fans from their music, from each other, and ultimately from themselves. The artificial intelligence of CRM precludes unanticipated exposure to new artists and genres that fall outside of its constructed affinities, cultivating a narrow exposure to genres. Fan discourse about music is supplanted and even squelched. As music services replace social spaces for sharing music, the fan "scenes" that are vital to the creation of music start to die off.

ALTERNATIVES TO THE CELESTIAL JUKEBOX

In their various public attacks on music "theft," the Big Four present the sale of copy-protected files as their main way to make money online. But there are other ways. eMusic still sells access to MP3 files without using DRM. New revenue models are also possible. For example, online music catalogs might be legally available only through compulsory license fees. Downloaders would pay these fees on ISP subscriptions or on the purchase of media devices; the federal government through Congress or the Copyright Office would set the rates.[15] The pool of money thus generated could be divided among copyright holders according to the frequency with which their properties were streamed or downloaded. Such an arrangement is already in place for radio stations, which pay a blanket license fee for the right to broadcast most recordings. However, this system would raise the price of Internet access. It would require Congress to pass legislation that would authorize file sharing and set up a system for taxing ISPs and distributing the proceeds to labels, music publishers, and other copyright holders. Perhaps most significantly, it could also provide less revenue to copyright holders in the short run. Given these institutional obstacles, a compulsory license for online music use remains an unlikely scenario at present.

Meanwhile, various forces continue to prevent the cultural commons from experiencing total lockdown by the culture industries: the Darknet, the various new disruptive technologies that widen the online bottlenecks, and the activities of public-interest movements that challenge the Celestial Jukebox model for commercialization of culture. Privacy activists, alternative-media promoters, open-source and free software developers, consumer advocates, and independent artists are converging to create and maintain real alternatives and choices to the majors and their panoptic technology. The collaborations among media-activist groups comprising the National Conference on Media Reform (NCMR) in North America delivered results in 2003 when, stemming from grassroots campaigning by NCMR, the U.S. Congress took the unusual step of overturning FCC rules on media ownership. The NCMR also emphasizes public access to broadband cable and wireless Internet infrastructures, cable- and satellite-TV distribution systems, and intellectual-property-rights reforms suitable for cyberliberties in a digital democracy.

"Hacktivism" focuses more intently than ever on preserving the anonymity of P2P exchanges:

> Three years after the [RIAA's] lawyers succeeded in shutting down the Napster file-trading service, the music industry's jihad against unauthorized music distribution is reaping an unintended consequence: better, easier-to-use software for exchanging data securely—and even anonymously—on the Internet.[16]

A "pay peer-to-peer" hybrid is developing as an alternative to both the Darknet and the Celestial Jukebox. The Mashboxx project is intended to blend the usability of ad hoc networks with a DRM technology called Snocap (created, ironically, by Napster founder Shawn Fanning). Weedshare encourages viral marketing by offering paid services and rewarding sharing. CDBaby organizes independent artists into an online collective for online distribution. Downhill Battle organizes musicians and fans in anticipation of, and in response to, RIAA practices. The outcome of these experiments remains to be seen, as rapid organizational, technological, and legal changes accompany the structural transformations of an old-line manufacturing industry.

Participation as an artist or fan within the trusted system is highly restricted, but participation *outside* the Jukebox framework continues to flourish. An alternative Jukebox, while vulnerable to industry broadsides and feints, may improve the availability and marketing of the music (nearly one-quarter of the total commercial supply) that is not controlled by the Big Four. The Independent Online Distribution Alliance (IODA) seeks to foster this alternative; it provides independent musicians with a menu of services, including collective bargaining with the Big Four, digital-music encoding, royalty distribution, marketing and promotion, and assistance with licensing. The viability of all these alternatives will be affected by court interpretations of the Grokster case, decided by the U.S. Supreme Court in 2005, which reinforced infringement standards established in the Napster case and increased the burden of proof of noninfringement for implicated parties in RIAA disputes.

Consequences of the Jukebox Culture

The Big Four's delayed entry into Internet distribution was the first of several calculated risks that the music industry took to ensure control over its new distribution platform:

> The music industry is just stalling through litigation until it can establish a standard, secure digital-encryption format, which is an essential step toward a global "pay-per-view" culture. This technocratic regime will be a severe threat to democracy and creativity around the world. . . . It involves the efforts of the content industries to create a "leakproof" sales and delivery system so they can offer all their products as streams of data triple-sealed by copyright, contract and digital locks. Then they can control access, use and ultimately the flow of ideas and expressions.[17]

In exchange for consumer "convenience," the Jukebox erodes our rights to privacy, fair use, and free speech. The extension and deepening of copyright protection, and the criminalization of unauthorized use of copyrighted material, hurts everyone seeking access to culture and knowledge in an information society. The simultaneous expansion of the rights of intellectual-property owners and contraction of the rights of the public strongly reinforces the asymmetrical power relationships institutionalized in the Celestial Jukebox between consumers and the content cartel. The Jukebox may promise more innovative music, more communities of interest for consumers, and lower prices for music; in fact, however, it gives us less (music in partial or damaged or disappearing files) and takes from us more (our privacy and our fair-use rights) than the old system. Opting out of the Jukebox in favor of alternative sources for music may be a better choice for many music fans, although those who continue using the Darknet face greater surveillance, shrinking catalogs, spoofed MP3 files, and other guerrilla-warfare tactics waged by agents of the Big Four.

Rather than a garden of abundance, the Celestial Jukebox offers a metered rationing of access to tiered levels of information, knowledge, and culture, based on the ability to pay repeatedly for goods that formerly could be purchased outright or copied for free. The entertainment industry places the onus of responsibility for the restrictive qualities of this new model on "pirates" whose "theft" will destroy the industry and cannot otherwise be thwarted. But their retaliatory measures are not equitable, and they damage critical institutions for a democratic civil society. As common cultural stakeholders, we have a collective responsibility, through politics, social movements, and cultural and technological innovation, to resist with creative alternatives the new media model built by the RIAA and the MPAA. Empowered citizens in an information society can ask for, and get, real antitrust regulation, compulsory licensing, proconsumer and proresearcher revisions to the DMCA, and universal service programs for online access to music and culture. At this point, civil society still affords some room for the public education, litigation, and lobbying that will be required in this effort, although the size of its space is shrinking as the Jukebox model expands.

Resistance to the Jukebox

Opposition to the Celestial Jukebox is developing overtly and covertly from a number of sides. The free-software community contributes software code not only for peer-to-peer networking but to other potentially disruptive technologies, such as the Mozilla Firefox browser, the free operating system Linux, and the free Apache Web server.[18] Ian Clarke, the founder and principal developer of Freenet, considers his work to be an expression of a libertarian political philosophy: "In the simplest terms possible . . . Freenet attempts to permit true freedom of speech. Copyright law attempts to prevent communication in some circumstances. And therefore, in order for Freenet to do its job successfully, it must prevent enforcement of copyright law."[19] Anonymity, free culture (or, in Clarke's term, "free speech"), privacy, access (fair use), and diversity remain core values among democratic activists in cyberspace. Software writers are joined in these values by culture jammers and "illegal" artists appropriating popular cultural symbols without permission. These artists include Tom Forsythe, a visual artist and Web publisher who was attacked by toy giant Mattel for publishing "Food Chain Barbie" on his website in 1999,[20] and DJ Dangermouse, who produces his "illegal art" by releasing copies of his live performances and studio music to the Internet. His *Gray Album*, mixing Jay-Z's *The Black Album* and the Beatles' *White Album*, provoked a DMCA-derived "takedown notice" by EMI. In response, online activists published the *Gray Album* en masse during "Gray Tuesday" in 2004.

These Internet activists are joined by public-interest groups devoted to cyberliberties and defense of "electronic frontiers," such as the Electronic Privacy Information Clearinghouse (EPIC), the Electronic Frontiers Foundation, the Free Software Foundation, Creative Commons, the Communication Rights in the Information Society (CRIS), the Internet Society, the Computer and Communications Industry Association, Computer Professionals for Social Responsibility, and the American Civil Liberties Union. Such groups seek to influence public policy at local, national, and global levels so as to preserve in cyberspace formal rights and liberties comparable to those enjoyed in real life during peacetime: free speech, privacy, market competition, open standards, open networks, and diversity in media and computing platforms. These groups make up the social and community grounding of resistance. All of these forces have the combined strength of a genuine global political movement. Collectively, they are ratcheting up the legal, technical, and cultural sophistication of their challenges. Although the copyright cartel represented by the Big Four would have us believe otherwise, the Celestial Jukebox is neither natural nor inevitable.

NOTES

1. Jon Pareles, "With a Click, a New Era of Music Dawns," *New York Times*, 15 Nov. 1998, 22.

2. Jeremy Rifkin, "Where Napster Has Gone, Others Will Follow," *Los Angeles Times*, 21 Aug. 2000, C9.

3. Simon Frith, "Look! Hear! The Uneasy Relationship of Music and Television," *Popular Music* 21, no. 3 (2002): 278.

4. Peter DiCola and Kristin Thomson, "Radio Deregulation: Has It Served Citizens and Musicians?" *The Future of Music Coalition Report*, 18 Nov. 2002, 65.

5. DiCola and Thomson, "Radio Deregulation," 66.

6. DiCola and Thomson, "Radio Deregulation," 64.

7. Pew Internet, "Reports: Technology and Media Use," 5 Dec. 2004, http://www.pewinternet.org/PPF/r/142/report_display.asp.

8. Greg Kot, "Who Stole Metallica's Money? Record Companies Are the Culprits, Not Napster Users," *Chicago Tribune*, 17 May 2000, 1(2).

9. Neil Strauss, "Ween Thrives on the Web," *New York Times*, 7 Aug. 2003, E3.

10. Steve Zisson, "Digital Hollywood: Subscription Models No Longer 'S' Word," *Webnoize News*, 18 May 2000, http://web.archive.org/web/20000619100823/http://news.webnoize.com/item.rs?ID=9077.

11. Future of Music Coalition, "iTunes and Digital Downloads: An Analysis," n.d., http://www.futureofmusic.org/itunes.cfm.

12. Eric Olsen, "Metallica Rethinks the Internet," *MSNBC*, 16 June 2003, http://opensourcemusic.com/content.asp?contentid=166.

13. Don Clark and Martin Peers, "Music Companies Are Hoping Downloads for Fees Can Prove as Popular as Free," *Wall Street Journal*, 20 June 2000, B1.

14. Michelle Goldberg, "Mood Radio," *San Francisco Bay Guardian*, 5 Nov. 2000, http://sfbg.com/noise/05/mood.html.

15. Christopher Stern, "Hill Takes Notice of Napster Legal Fray," *Washington Post*, 16 Feb. 2001, E3.

16. Matthew Fordahl, "21st-Century Speakeasies: Internet Evolves in Wake of Music-Swapping Suits," *Ottawa Citizen*, 9 Oct. 2003, E2.

17. Siva Vaidhyanathan, "MP3: It's Only Rock and Roll and the Kids Are Alright," *The Nation*, 24 July 2000, 34.

18. Chris DiBona, Sam Ockman, and Mark Stone, eds., *Open Sources: Voices from the Open Source Revolution* (Beijing: Sebastopol O'Reilly and Associates, 1999).

19. Kevin Featherly, "Freenet: Will It Smash Copyright Law?" *Linux Today*, 21 Mar. 2001, http://linuxtoday.com/news_story.php3?ltsn=2001-03-21-021-04-PS-CY.

20. Illegal Art, "Illegal Art: Freedom of Expression in the Corporate Age," *Stay Free!* n.d., http://www.illegal-art.org/index.html.

Selected Bibliography

"Entertainment Revenue Reaches $53 Billion in '03." *Entertainment Marketing Letter* 17, no. 3 (1 Feb. 2004).

"Metrics." *EContent* 25, no. 11 (Nov. 2002): 15.

"Who Owns What: Time Warner." *Columbia Journalism Review*. 11 Aug. 2004. http://www.cjr.org/tools/owners/timewarner.asp.

Adams, Henry. *The Education of Henry Adams*. 1907. Reprint, New York: Vintage Books, 1990.

Albini, Steve. "The Problem with Music." http://www.negativland.com/albini.html.

Alderman, John. *Sonic Boom: Napster, P2P, and the Battle for the Future of Music*. Cambridge, MA: Perseus Press, 2001.

Anonymous. "Thaddeus Cahill's 'Dynamophone/Telharmonium.'" http://www.obsolete.com/120_years/machines/telharmonium/.

Aufderheide, Patricia. "Competition and Commons: The Public Interest in and after the AOL-Time Warner Merger." *Journal of Broadcasting and Electronic Media* 46, no. 4 (2002): 515–31.

Bagdikian, Ben. *The Media Monopoly*. 6th ed. New York: Beacon Press, 2000.

Biddle, Peter, Paul England, Marcus Peinado, and Bryan Willman. "The Darknet and the Future of Content Distribution." Paper presented at the 2002 ACM Workshop on Digital Rights Management, Washington, D.C., 18 Nov. 2002. http://crypto.stanford.edu/DRM2002/darknet5.doc.

Brealey, Richard A., and Stewart C. Myers. *Principles of Corporate Finance*. 7th ed. Boston: McGraw-Hill Irwin, 2002.

Bricklin, Dan. "The Cornucopia of the Commons." In *Peer-to-Peer: Harnessing the Benefits of a Disruptive Technology*, edited by Andy Oram, 59–63. Cambridge: O'Reilly & Associates, 2001.

Burkart, Patrick. "Loose Integration in Markets for Popular Music." *Popular Music & Society*, 28, no. 4 (2005): 489–500.

Burkart, Patrick, and Tom McCourt. "Infrastructure for the Celestial Jukebox," *Popular Music* 23 (2004): 349–62.

Burnett, R. "The Implications of Ownership Changes on Concentration and Diversity in the Phonogram Industry." *Communication Research*, 19 (1992): 749–69.

Camp, L. Jean. *Trust and Risk in Internet Commerce*. Cambridge, MA: MIT Press, 2000.

Carroll, Amy, Mario Juarez, Julia Polk, and Tony Leininger. "Microsoft 'Palladium': A Business Overview." Microsoft Corporation, Aug. 2002.

Chan-Olmstead, Sylvia, and Byeng-Hee Chang. "Diversification Strategy of Global Media Conglomerates: Examining Its Patterns and Determinants." *Journal of Media Economics* 16, no. 4 (2003): 213–33.

Chong, Rachelle B. "FCC Commissioner Chong Hails Passage of New Telecom Bill." Federal Communications Commission, 1 Feb. 1996. http://www.fcc.gov/Speeches/Chong/sprbc602.txt.

Cooper, Jon, and Daniel M. Harrison. "The Social Organization of Audio Piracy on the Internet." *Media Culture & Society* 23 (2001): 71–89.

DiBona, Chris, Sam Ockman, and Mark Stone, eds. *Open Sources: Voices from the Open Source Revolution*. Beijing: Sebastopol O'Reilly and Associates, 1998.

DiCola, Peter, and Kristin Thomson. *Radio Deregulation: Has It Served Citizens and Musicians?* The Future of Music Coalition Report. 18 Nov. 2002.

Dolfsma, Wilfred. "How Will the Music Industry Weather the Globalization Storm?" *First Monday*, 5 May 2000. http://firstmonday.org/issues/issue5_5/dolfsma/index.html.

Ennis, Philip H. *The Seventh Stream: The Emergence of Rock'n'Roll in American Popular Music*. Hanover: Wesleyan University Press, 1992.

European Commission. "Commission Decides Not to Oppose Recorded Music JV between Sony and Bertelsmann." IP/04/959, Brussels, 20 July 2004. http://europa.eu.int/rapid/pressReleasesAction.do?reference=IP/04/959&format=HTML&aged=0&language=EN&guiLanguage=en.

Feldman, Tony. *Introduction to Digital Media*. New York: Routledge, 1997.

Fordahl, Matthew. "Internet Evolves in Wake of Music-Swapping Suits." *USA Today*, 5 Oct. 2003. http://www.usatoday.com/tech/webguide/internetlife/2003-10-05-internet-underground_x.htm.

Frith, Simon. "Video Pop: Picking Up the Pieces." In *Facing the Music*, ed. Simon Frith, 88–130. New York: Pantheon, 1988.

———. "Music Industry Research: Where Now? Where Next? Notes from Britain." *Popular Music* 19, no. 3 (2000): 387–93.

——. "Look! Hear! The Uneasy Relationship of Music and Television." *Popular Music* 21, no. 3 (2002): 277–90.

Future of Music Coalition. "iTunes and Digital Downloads: An Analysis." n.d. http://www.futureofmusic.org/itunes.cfm.

Gandy, Oscar H., Jr. *The Panoptic Sort: A Political Economy of Personal Information*. Boulder: Westview Press, 1993.

Garofalo, Reebee. "From Music Publishing to MP3: Music and Industry in the Twentieth Century." *American Music* 17, no. 3 (1999): 318–53.

Gartner G2 and the Berkman Center for Internet & Society. *Copyright and Digital Media in a Post-Napster World*. Harvard Law School, 2003. http://cyber.law.harvard.edu/home/2003-05.

Gershon, Richard A. "The Transnational Media Corporation: Environmental Scanning and Strategy Formulation." *The Journal of Media Economics* 13, no. 2 (2000): 81–101.

Goldstein, Paul. *Copyright's Highway: The Law and Lore of Copyright from Gutenberg to the Celestial Jukebox*. New York: Hill and Wang, 1994.

Gore, Al. "Remarks Prepared for Delivery by Vice President Al Gore." Royce Hall, UCLA, Los Angeles, California, 11 Jan. 1994. http://www.ibiblio.org/icky/speech2.html.

Grokster and Electronic Frontiers Foundation. Brief in opposition. *Metro-Goldwyn-Mayer Studios Inc., et al., Petitioners, v. Grokster, Ltd., et al., Respondents*, on petition for writ of certiorari to the United States Court of Appeals for the Ninth Circuit. 8 Nov. 2004. http://www.eff.org/IP/P2P/MGM_v_Grokster/20041108_Final_Brief.pdf.

Hayward, Philip. "Enterprise on the New Frontier: Music, Industry and the Internet." *Convergence* 1, no. 2 (1995): 29–44.

Hazlett, Thomas W. "Rent-Seeking in the Telco/Cable Cross-Ownership Controversy." *Telecommunications Policy* 14, no. 5 (Oct. 1990): 425–33.

Hillman, Amy. "Napster: Catalyst for a New Industry or Just Another Dot-com? (The Ivey Case Study[C]): Part 3 of 3." *Ivey Business Journal* 66, no. 3 (Jan. 2002).

Hugenholtz, P. Bernt. "The Great Copyright Robbery: Rights Allocation in a Digital Environment." Paper presented at a Free Information Ecology in a Digital Environment Conference, New York University School of Law, 31 Mar.–2 Apr. 2000.

Hundt, Reed. *You Say You Want a Revolution: A Story of Information Age Politics*. New Haven, CT: Yale University Press, 2000.

Illegal Art. "Illegal Art: Freedom of Expression in the Corporate Age." *Stay Free!* n.d. http://www.illegal-art.org/index.html.

Jones, Steve. "Music That Moves: Popular Music, Distribution and Network Technologies." *Cultural Studies* 16, no. 2 (2002): 213–32.

Jossie, Frank. "Net Perceptions Hits the Big Time." Interactive Pioneers, n.d. http://www.interactive-pioneers.org/net_perceptions.html.

Krasilovsky, M. William, and Sidney Shemel. *This Business of Music: The Definitive Guide to the Music Industry*. 8th ed. New York: Billboard Books, 2000.

Lanza, Joseph. *Elevator Music: A Surreal History of Muzak, Easy-Listening, and Other Moodsong*. New York: St. Martin's Press, 1994.

Latonero, Mark. "Survey of MP3 Usage: Report on a University Consumption Community." Annenberg School of Communication, University of Southern California, June 2000.

Lessig, Lawrence. *Code and Other Laws of Cyberspace*. New York: Basic Books, 1999.

———. *The Future of Ideas: The Fate of the Commons in a Connected World*. New York: Random House, 2002.

Leyshon, Andrew. "Scary Monsters? Software Formats, Peer-to-Peer Networks, and the Spectre of the Gift." *Environment and Planning D: Society and Space* 21 (2003).

Leyshon, Andrew, David Matless, and George Revill, eds. *The Place of Music*. New York: The Guilford Press, 1998.

Lopes, Paul D. "Innovation and Diversity in the Popular Music Industry, 1969 to 1990." *American Sociological Review* 57 (1992): 56–71.

Lovering, John. "The Global Music Industry: Contradictions in the Commodification of the Sublime." In *The Place of Music*, edited by Andrew Leyshon, David Matless, and George Revill. New York: The Guilford Press, 1998.

Marcus, Adam David. "The Celestial Jukebox Revisited: Best Practices and Copyright Law Revisions for Subscription-Based On-line Music Services." Unpublished paper, 2003. http://tprc.org/papers/2003/220/TPRC_paper-Adam_Marcus.pdf.

McChesney, Robert W. *Corporate Media and the Threat to Democracy*. New York: Seven Stories Press, 1986.

McCourt, Tom, and Patrick Burkart. "When Creators, Corporations, and Consumers Collide: Napster and the Development of On-Line Music Distribution." *Media, Culture and Society* 25, no. 3 (2003).

McLeod, Kembrew. *Freedom of Expression: Overzealous Copyright Bozos and Other Enemies of Creativity*. New York: Doubleday, 2005.

Meyers, Cynthia. "Entertainment Industry Integration Strategies." Paper presented at the Annual Institute for International Studies and Training Conference, Tokyo, Japan, 14 Feb. 2002. http://www.iist.or.jp/wf/magazine/0068/0068_E.html.

Millard, Andre. *America on Record: A History of Recorded Sound*. New York: Cambridge University Press, 1996.

Minar, Nelson, and Marc Hedlund. "A Network of Peers: Peer-to-Peer Models through the History of the Internet." In *Peer-to-Peer: Harnessing the Benefits of a Disruptive Technology*, edited by Andy Oram. Sebastopol, CA: O'Reilly & Associates, 2001.

Mosco, Vincent. *The Pay-Per Society: Computers and Communication in the Information Age*. Mahwah, NJ: Ablex, 1990.

Naughton, Russell. "Adventures in Cybersound." n.d. http://www.acmi.net.au/AIC/TV_TL_COMP_1.html.

Negus, Keith. *Music Genres and Corporate Cultures*. London: Routledge, 1999.

Oberholzer-Gee, Felix, and Koleman Strumpf. "The Effect of File Sharing on Record Sales: An Empirical Analysis." Mar. 2004. http://www.unc.edu/~cigar/papers/FileSharing_March2004.pdf.

Pate, R. Hewitt. Statement by Assistant Attorney General R. Hewett Pate regarding the closing of the digital music investigation, U.S. Department of Justice, 23 Dec. 2003. http://www.usdoj.gov/atr/public/press_releases/2003/201946.pdf.

Peters, Jim. "Copy-Protected CDs." Nov. 2001. http://www.ukcdr.org/issues/cd/.

Peterson, Richard A., and David Berger. "Cycles in Symbol Production: The Case of Popular Music." *American Sociological Review* 40 (1975): 158–73.

Pew Internet. "Reports: Technology and Media Use." 5 Dec. 2004. http://www.pewinternet.org/PPF/r/142/report_display.asp.

Recording Industry Association of America. "RIAA Issues 2004 Year-End Shipment Numbers." 21 Mar. 2005. http://www.riaa.com/news/newsletter/032105.asp.

Reidl, John, Joseph Konstan, and Eric Voorman. *Word of Mouse: The Marketing Power of Collaborative Filtering*. New York: Warner Books, 2002.

Rheingold, Howard. *Smart Mobs: The Next Social Revolution*. Cambridge, MA: Perseus Publishing, 2002.

Robinson, Deanna Campbell, Elizabeth Buck, and Marlene Cuthbert. *Music at the Margins: Popular Music and Global Cultural Diversity*. Newbury Park, CA: Sage, 1991.

Rosenblatt, Bill, and Gail Dykstra. "Integrating Content Management with Digital Rights Management: Imperatives and Opportunities for Digital Content Lifecycles." Giantsteps Media Technology Strategies. 14 May 2003. http://www.drmwatch.com/resources/whitepapers/article.php/3112011.

Rothenbuhler, Eric W., and John Dimmick. "Popular Music: Concentration and Diversity in the Industry, 1974–1980." *Journal of Communication* 32, no. 1 (1982): 143–49.

Rothenbuhler, Eric W., and Tom McCourt. "The Economics of the Recording Industry." In *Media Economics: Theory and Practice*, edited by Alison Alexander, et al. 3rd ed. Mahwah, NJ: Lawrence Erlbaum, 2004.

Sanjek, Russell. *American Popular Music and Its Business, Vol. 3: From 1900 to 1984*. New York: Oxford University Press, 1988.

Sanjek, Russell, and David Sanjek. *American Popular Music Business in the 20th Century*. New York: Oxford University Press, 1991.

Sassen, Saskia. "Digital Networks and Power." In *Spaces of Culture: City—Nation—World*, edited by Mike Featherstone and Scott Lash. Thousand Oaks, CA: Sage, 1999.

Schatz, Tom. "The Return of the Hollywood Studio System." In *Conglomerates and the Media*, edited by Patricia Aufderheide, et al. W. W. Norton: New York, 1997.

Schaumann, Neils. "Copyright Infringement and Peer-to-Peer Technology." *William Mitchell Law Review* 28, no. 3 (2002): 1001–46.

Segrave, Kerry. *Jukeboxes: An American Social History*. Jefferson, NC: McFarland & Company, 2002.

Shirky, Clay. "Listening to Napster." In *Peer-to-Peer: Harnessing the Benefits of a Disruptive Technology*, edited by Andy Oram. Sebastopol, CA: O'Reilly & Associates, 2001.

Skaflestad, Odd Arild, and Nina Kaurel. "Peer-to-Peer Networking: Configuring Issues and Distributed Processing." Paper presented at an NTNU conference, 2001. http://www.item.ntnu.no/fag/SIE50AC/P2P.pdf.

Stallabrass, Julian. *Gargantua: Manufactured Mass Culture*. New York: Verso, 1996.

Standage, Tom. *The Victorian Internet: The Remarkable Story of the Telegraph and the Nineteenth Century's On-Line Pioneers*. New York: Walker, 1998.

Thompson, Mozelle W. "Statement of Commissioner Mozelle W. Thompson." Sony Corporation of America/Bertelsmann Music Group Joint Venture, File No. 041-0054. (2004). http://www.ftc.gov/os/caselist/0410054/040728mwtstmnt0410054.pdf (2004).

United States Court of Appeals for the Seventh Circuit. (2003). No. 02-4125 IN RE: AIMSTER COPYRIGHT LITIGATION. APPEAL OF: JOHN DEEP, Defendant. Appeal from the United States District Court for the Northern District of Illinois, Eastern Division. No. 01 C 8933—Marvin E. Aspen, *Judge*. ARGUED JUNE 4, 2003—DECIDED JUNE 30, 2003. http://www.riaa.com/news/newsletter/pdf/aimster20030630.pdf.

U.S. Department of State. "Copyright in an Electronic Age." USInfo publication, n.d. http://usinfo.state.gov/products/pubs/intelprp/copyrt.htm.

Vaidhyanathan, Siva. *Copyrights and Copywrongs: The Rise of Intellectual Property and How it Threatens Creativity*. New York: New York University Press, 2001.

Vignoli, Fabio. "Digital Music Interaction Concepts: A User Study." Paper presented at the Fifth International Conference on Music Information Retrieval, Barcelona, Spain, 10–14 Oct. 2004. http://ismir2004.ismir.net/proceedings/p075-page-415-paper152.pdf.

Vivendi Universal. Form 20-F of Annual Report Pursuant to Section 13 or 15(d) of the Securities Exchange Act of 1934. 2004.

Vivendi Universal. "Vivendi Universal Preliminary Supplemental Revenues Information." 2003. http://finance.vivendiuniversal.com/finance/documents/financial/annual-release/2002/PR100203Charts.pdf.

Vogel, Harold L. *Entertainment Industry Economics*. 4th ed. Cambridge, UK: Cambridge University Press, 1998.

Webb, Jon. "Digital Rights Management or DRM—Does (It) Really Matter?" White paper of the International Association of Scientific, Technical and Medical Publishers, 2004. http://www.stm-assoc.org/annualreport/00drm.html.

Wellman, Barry. "Designing the Internet for a Networked Society." *Communications of the ACM* 45, no. 5 (May 2002): 96.

Willie, Boxcar. "Boxcar Willie: The World's Favorite Hobo." http://www.boxcarwillie.com/.

Winston, Brian. *Media Technology and Society: A History from the Telegraph to the Internet*. New York: Routledge, 1998.

World Intellectual Property Organization. "About WIPO." http://www.wipo.int/about-wipo/en/overview.html.

Yeager, Shawn. Comment posted to Pho listserve. 7 July 2003. http://www.pholist.org.

Zvonar, Richard. "Spatial Music." Unpublished essay, 1999. http://www.zvonar.com/writing/spatial_music/History.html.

Index